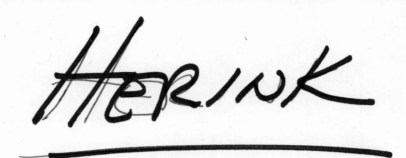

Herzog

KANBAN
JUST-IN-TIME
AT TOYOTA

KANBAN
JUST-IN-TIME
AT TOYOTA

Management Begins
at the Workplace

Edited by the Japan Management Association

Translated by David J. Lu

Forward by Norman Bodek, President, Productivity, Inc.

Productivity Press
Stamford, Connecticut   Cambridge, Massachusetts

Originally published as *Toyota no genba kanri: kanban hōshiki no tadashii susumekata*, copyright © 1985 Japan Management Association, Tokyo.

English translation copyright © 1986 by Productivity, Inc.

Productivity Press   or   Productivity, Inc.
P.O. Box 814   P.O. Box 16722
Cambridge, MA 02238   Stamford, CT 06905
(617) 497-5146   (203) 967-3500

Library of Congress Catalog Card Number: 85-63443
ISBN: 0-915299-08-9

Cover design: Russell Funkhouser
Composition by Rudra Press, Cambridge, MA
Cover printed by Phoenix Color Corp.
Printed and bound by Arcata Graphics/Fairfield
Printed in the United States of America

Library of Congress Cataloging-in-Publication Data

Toyota no genba kanri.   English.
    Kanban and just-in-time at Toyota; management begins at the workplace.

    Translation of: Toyota no genba kanri: kanban hōshiki no tadashii susumekata.
    1. Production control.   2. Toyota Jidōsha
Kōgyō Kabushiki Kaisha.   I. Lu, David J. (David John)
II. Keizai Dōyūkai.   III. Title.
TS157.T6913   1986   658.5   85-63443
ISBN 0-915299-08-9

Publisher's Preface

Have you ever envisioned yourself on a safari in Africa or a voyage to the exotic Orient, looking for hidden treasure, diamond mines or priceless antiques — in short, for a once-in-a-lifetime discovery that would cure all your ills and take care of you forever? With this discovery, you imagine, you wouldn't have to work or worry for the rest of your life.

In our search for knowledge we often look for "the answer," the one thing that will solve all of our problems. Naive as it may sound, I believe that in this marvelous book you will find that one answer — the "magic potion" that will make your company a world-class competitor and make you personally very wealthy.

The truth discovered by Mr. Taiichi Ohno, Vice President of Toyota Motors and father of the kanban and Just-In-Time concepts, was that improvement must never stop. It is based on the samurai tradition that a warrior (a manager) never stops perfecting his style (improving his managerial ability), and never stops polishing his sword (improving the process and the product).

The secret to success, then, is the never-ending search for better ways to improve the productivity of the process and the quality of the product. You never stop and say, "Ah, now I've made it. I have the answer." The answer lies only in the quest for more and more knowledge, for greater awareness to find newer and simpler solutions, for ways to be more innovative. Within the pages of this book you will find the very clear understanding and direction given by Mr. Ohno to his managers on how to continually improve.

Recently, a group of American travelers to Japan asked a group of Japanese managers which company leads Japan in quality and productivity. The answer was unanimous: "Toyota, because they never stop making improvements and refinements to their system."

Last year Toyota received an average of 40 new improvement ideas per worker. Imagine running a company where every person is actively involved in making suggestions to improve his or her job, to improve the company. The thought alone is mind boggling.

The concepts explained so clearly in this book are being applied throughout Japan and by numerous American companies. As a matter of fact, I spoke just recently with a manager at Omark Industries. His company, in 1983, bought 500 copies of Shigeo Shingo's book *Study of Toyota Production System*. (Mr. Shingo worked with Mr. Ohno to develop the Toyota system.) They asked all their managers to read the book and begin to apply the ideas to their plant. The results after only a few years have been amazing:

- Lead time on one product was reduced from 12 weeks to 4 days;
- Setup time on a large blanking press was reduced from 8 hours to 1 minute and 4 seconds;
- Work in process plant-wide has been reduced 50 percent;
- Factory floor space was opened up 30 to 40 percent in every one of their plants.

As you read this book, you will begin to see clearly the real magic in the Toyota system. I recommend that you read the book in study groups with other people in your company. Ask each other how you can use the information to improve your company. Above all, don't make the mistake of thinking that the ideas "don't apply here." The concepts are applicable to repetitive manufacturing, to job shops, to process industries, to virtually any manufacturing company. And through careful understanding, you will begin to see how the ideas can also be applied to the office.

Mr. Ohno asks us to reexamine the way we think and to begin focusing on ways to eliminate waste and continually find ways to improve the quality of the process and the product.

I hope you enjoy the adventure of this book, as I have. I want to express my appreciation to all those involved in producing this book: Mr. Taiichi Ohno, the inventor; the managers at Toyota who allowed their words to be taped and transcribed; the Japan Management Association, who compiled the original material and first published it. Many thanks also to David Lu, the translator, whose knowledge of the subject gives an added dimension to the material. Patty Slote oversaw all editorial and production details; Cheryl Berling edited the manuscript; Marie Kascus prepared the index; Russ Funkhouser designed the cover; and Nanette Redmond, Laura Santi, Leslie Goldstein and Ruth Knight of Rudra Press typeset and pasted up the book.

Norman Bodek

Translator's Preface

I have visited Toyota's assembly plant a couple of times, and each time I've been impressed by the relaxed manner in which its workers go about their work. "Work smarter, not harder," says Dr. W. Edwards Deming, the Dean of quality control activities, in whose honor a prestigious Japanese prize is named. The Toyota workers seem to typify that spirit while producing world-class quality cars.

For Toyota to reach this stage of development and sophistication, it had to go through a rather rocky road. In 1949, it was at the brink of bankruptcy, and the consensus in Japan in those early postwar years was that Japan was not fit to produce passenger cars. America was then at least eight times as efficient as Japan in car production. Nevertheless, a determined Toyota president, Kiichiro Toyoda, issued a challenge to Toyota to catch up with the United States within three years.

"America's productivity is probably greater than eight or nine times. However, I cannot believe that Americans are physically exerting ten times as much energy as the Japanese," Taiichi Ohno thought to himself when he heard Toyoda's challenge. "It's probably that the Japanese are very wasteful in their production system." Ohno's solution was to organize a production system which was dedicated to total elimination of waste. The Toyota production system had its genesis in this thought process.

The Toyota system brings inventory down to close to nothing through the application of the *just-in-time system*. To make this system function smoothly, the subsequent process must come to the preceding process to withdraw parts and materials at the time needed and in the quantity needed.

The preceding process must produce only the exact quantity withdrawn by the subsequent process. To indicate which process needs what and to allow various processes to communicate with each other efficiently, the *kanban system* was created. The company's production plan is given only to the final assembly line. When it goes to the preceding process to withdraw parts and materials, it establishes a chain of communication with each preceding process, and every process automatically knows how much and when to produce the parts and materials it is assigned to produce.

Another characteristic of the Toyota system is *automation with a human touch*. Machines are taught to do what people can do, and when they produce defectives, they are taught to stop automatically. In his carefully written work, *Toyota Production System: Practical Approach to Production Management* (1983), Professor Yasuhiro Monden describes this phenomenon with the term *autonomation*. Yes, machines are taught to be autonomous, but somehow this term does not fully convey the meaning which Mr. Ohno intended. The original Japanese term is *ninben no tsuita jidoka*. The word *jidoka* means automation. Here the key to our understanding can be found in the first word, *ninben*, which is a radical added to the Chinese or Japanese character to make it represent the action of a human being. Ohno added this *ninben* to the second character of the word *jidoka* (automation), which standing alone would have simply meant "to move," but with the addition of the *ninben* meant "to work." And the entire phrase would assume a different meaning suggesting that the machines are endowed with human intelligence and touch. Hence my preferred translation of "automation with a human touch."

Machines with these human capabilities stop automatically when abnormalities occur. The art of management is greatly affected by this. As long as the machines function normally, no attendance by the workers is required. Only when a machine stops, or an abnormality occurs, is there a need for a worker to be present. Thus it becomes possible for one worker to handle a number of machines sequentially. *Visual control* also becomes possible when hidden defects or abnormalities are made apparent.

The Toyota system is rational and cost effective. Along with the institution of the one-shot exchange of dies, it allows production of many different styles and types of cars in small lots. The Toyota system did not receive the accolades it deserved in the mid-1950s through the 1960s when Japan experienced an annual double-digit percentage increase in its gross national product. Only when the first oil shock of 1973 exposed the limit to Japan's economic expansion was close attention paid by other industries to the utility and potentials suggested by the Toyota system. In a period of slower growth, it provides a role model for other sectors to follow, showing that profits can still be made by changing the methods of production.

The book that is translated here is an outgrowth of a series of seminars conducted by the Japan Management Association in the mid-1970s to teach others about the Toyota production system. The seminar instructors were Taiichi Ohno, then an executive vice president, and staff members of the production control division of Toyota Motors. The instructional text written by the Toyota staff for these seminars serves as the basis of this book, which has been edited and revised by the editorial writers of the Japan Management Association. It was first published in 1978. By the summer of 1985 it had reached its 35th printing and become one of the best-read management books in Japan. Its appeal comes from its easy to understand style and from its homespun wisdom found both in the text and in the sayings of Ohno interspersed with the text. As companies, especially small and medium-sized ones, groped for ways to survive under difficult economic and social conditions, the Toyota method of work-place management, as shown in this book, gave them new direction and encouragement.

Can the Toyota production system be applied with benefit in the United States? The answer is definitely in the affirmative. This is so because Americans, like the Japanese, are rational people who can select and adopt the most suitable system for their own society, and the very rationality of the Toyota system should appeal to the American mind. The Toyota system can teach us how to eliminate wastes which are often not apparent to us. It is a system that demands that employees do their best, but does not overwork them. The system establishes itself as a friend of the workers and not their adversary. Implicit in the system is the philosophy of respect for humanity. The sense of trust created between management and the workers can promote efficiency and at the same time a relaxed feeling.

We may or may not transplant, graft or reject outright the just-in-time system for our manufacturing industries. But it is good to remember that Mr. Ohno received the inspiration for this system by observing the working of an American supermarket. Food items picked up by the customers were replaced "just in time" on the shelf for the next round of customers. If an American supermarket can inspire a major Japanese automaker to excel in its production method, so can the latter help the Americans stage a comeback to regain their industrial prominence.

David J. Lu

Contents

Figures

KANBAN
JUST-IN-TIME
AT TOYOTA

1

The Source of Profit Is In the Manufacturing Process

Commercial Profit and Manufacturing Profit

In 1976 and 1977 — shortly after the first oil shock — when Toyota Motors registered profits of ¥182.2 billion ($597.4 million)* and ¥210 billion ($716.7 million) respectively, the company was criticized for making too much money.

For a company to succeed, making money is actually a precondition or a goal, irrespective of the industry one is in. Now, what do we mean by the phrase "making money"?

In commercial enterprises, the selling price is set by adding a certain margin over the purchase price. To make money means "to buy cheaply and to sell dearly." Thus "making money" usually conveys a negative image, and some newspapers would even write articles condemning companies that make too much money as being engaged in anti-social activities. They reason that the companies make money by buying cheaply and selling dearly, making the consumers pay the difference.

In manufacturing, is money made by buying raw materials and parts cheaply and selling finished products at a higher price, as done in the commerical sector?

Does it mean that Toyota can somehow buy steel plates cheaper than any other car maker? Does it mean that there are suppliers

---

* Based on the exchange rates for those two years.

who are willing to sell parts at a lower price to Toyota? No, that is not the case. Can Toyota command a higher price by the use of its brand name? The mere fact that the Toyota name is on the car does not mean that it can automatically command a price $1000 higher than any other car.

Toyota buys its raw materials, processed materials, parts, electricity and water at the prevailing market value. The price of its products is also governed by the rules of the same marketplace. If Toyota should put an unreasonably high price on its cars, it would put an end to its sales drive everywhere.

This is not confined to Toyota. All manufacturers share the same marketing consideration. The manufacturing industries derive their profit from the added value obtained through the process of manufacturing. Therefore, manufacturing industries and commercial enterprises cannot make money in the same way.

We Cannot Be Guided by Cost Alone

If profit is expressed in terms of a margin obtained by selling at a higher price than the purchase price, or by selling products above the manufacturing cost, it can then be summarized in the following equation:

$$\text{Profit} = \text{Selling price} - \text{Cost}$$

On the other hand, if one wishes to take into account the purchase price and manufacturing cost before adding profit, another equation may be established as follows:

$$\text{Selling price} = \text{Cost} + \text{Profit}$$

When the two equations are expressed in numbers, they may be the same, but at Toyota, we do not use the "selling price = cost + profit" formula.

The so-called cost principle states that inasmuch as it costs so much to manufacture a certain product, a just amount of profit must be added to it to arrive at the selling price. Thus it becomes "selling price = cost + profit." If we were to insist on abiding by this cost principle, we would have to say to ourselves: "Well, we cannot help it if this product costs so much to make. We have to be able to make this much money out of it." This would mean that every cost would have to be borne by the consumer. We cannot afford to take this attitude in this age of intense competition. Even if we wanted to, we could not use this formula.

Returning to the first equation, it is stated that profit is the balance after subtracting cost from the selling price (profit = selling price − cost). As discussed earlier, the price of a car is determined generally by the marketplace. Thus in order to make a profit, the only recourse left to us is to *lower the cost as much as possible*. Herein lies the source of our profit.

SAYINGS OF OHNO

Don't confuse "value" with "price."

When a consumer buys a product, he does so because that product has a certain value to him.

The cost is up, so you raise your price! Don't take such an easy way out. It cannot be done. If you raise your price but the value remains the same, you will quickly lose your customer.

True Cost Is the Size of a Plum Seed

Cost can be interpreted in many different ways. Cost consists of many elements, such as personnel cost, raw materials cost, cost of oil, cost of electricity, cost of land, cost of buildings and cost of equipment. Some people may add all of these costs and obtain a total and say that it costs this much to manufacture a certain product. But is this the true cost? No, when one considers it carefully, what emerges is that the total just obtained does not seem to reflect the true cost at all.

The expression "true cost" may sound odd. But there is a notion that in making a passenger car the true personnel cost is about this much, and that only a certain amount of the cost of materials is sufficient. That is an approximation of the true cost.

Let us now take the personnel cost as an example. In order to make any given product, a worker must work a requisite number of hours to process a certain amount of material needed for the day.

That is close to the true cost. But suppose the worker processes those materials needed for tomorrow and the day after tomorrow?

The excess materials which are manufactured, if kept in the same workplace, will hinder the orderly functioning of the workplace. So they are shipped somewhere. This means that a process called *shipping* has to be created, and a need for a storage place also arises. Furthermore, someone has to count and rearrange these materials in the name of *management*. If the number increases, slips will be needed to show that certain items are placed in storage and certain items are removed from storage. Next comes the need for storage clerks, and then workers monitoring various processes.... Just because someone has overproduced, there is created a need for an unlimited amount of work and additional personnel.

Those people who are engaged in these newly created tasks must be paid, and that cost is counted as part of the personnel cost. In the end, their salaries and wages become part of the cost of that product.

The same thing can be said about the materials cost. If you have just enough materials for today's work, your day's work can run smoothly. And you may keep a ten-day supply, for the sake of your suppliers. That, of course, is more than sufficient. But in many companies, when inventory is taken, it is often discovered that they have supplies sufficient for one or two months lying idly in storage. It is not uncommon to have a supply for six months, which is not an acceptable condition.

Do not forget that these materials are already paid for. In addition to the materials cost, there is the interest charge. Furthermore, during storage, materials may be rusted, broken or disjoined to become odd pieces that cannot be used. In a more serious case, you may have design changes making the materials in storage obsolete. Then there are instances in which a shift in your sales may obviate the need for some materials. In any event, storage can create waste.

This waste, the cost of unused materials which are discarded, is also entered as the cost of materials by your accounting department, and it becomes part of the cost of that particular product.

In most instances, when people speak of *cost,* it is expressed in terms of a hybrid of just and unjust costs, and in the case of thelatter, it includes those portions of personnel and materials costs that are not really necessary in manufacturing a product.

In Toyota we have a saying: "The true cost is only the size of a plum seed." The trouble with most managers is that they have a penchant for bloating the plum seed into a huge grapefruit. They then shave off some unevenness from the rind and call it cost reduction. How wrong can they get?

Change Your Manufacturing Method, Lower Your Cost

At Toyota, we do not adhere to the so-called cost principle. Behind the cost principle lies the notion that "no matter how differently we manufacture our products, the cost remains the same." If it is proven correct that, regardless of the manufacturing methods, cost remains constant, then all industries must abide by the cost principle.

However, by changing its manufacturing method, a company can eliminate its personnel cost, which does not produce added value, and its materials cost, which pertains to those materials not used. By changing the manufacturing method, cost can be substantially reduced.

There is a Toyota subsidiary that makes metal-stamped parts and is located next to the Toyota headquarters. In 1973, it was at a standstill and all officers were replaced. Starting anew that year, employees did their very best, and two years later, in 1975, the company was completely recovered.

Today that company is a very profitable one. According to its president, one day an inspector from the National Tax Administration Agency came, ready to grill its officers. "Why is it that your company experienced a sizable deficit in 1973 when economy was at the height of a boom," asked the inspector, "and a good income in 1976 when there was a recession?"

The president's response was typical of Toyota: "That is what we call improvement and effort by the company." The inspector remained incredulous. At any rate, cost is changed by the method of manufacturing. Naturally the profit picture changes along with it, and the above provides a good example.

Production Technique and Manufacturing Technique

Today Toyota produces well over 200,000 units per month. In 1952, it took ten employees one month to produce one truck. In 1961, Toyota's monthly production was 10,000 units. There were 10,000 employees then, and it meant that every month one employee produced one passenger car. In the last couple of years, the monthly production ranged from 230,000 to 250,000 units, and we have 45,000 employees. This means that each employee is credited with making five passenger cars each month.

Toyota has a number of assembly plants overseas. There the number of processes required in assembling the same Corolla or Corona may be five to ten times that of Japan. For the same Toyota, depending on the time and place, there is this much difference.

How is this difference created? In part the difference in the production facility is responsible, but to a large degree, such a difference comes from the difference in the manufacturing methods.

For many years, we have thought about and improved our manufacturing method. That is what we call today the Toyota production system.

Two techniques are utilized in manufacturing. One is the production technique and the other is the manufacturing technique.

Simply stated, *production technique* means the technique needed to produce goods. Normally when the term *technique* is used, it refers to this production technique.

In contrast, *manufacturing technique* means the technique of expertly utilizing equipment, personnel, materials and parts. If we consider the production technique to be proper technique, conforming to established standards, then the manufacturing technique can be considered management technique, utilizing and synthesizing various methods. What we call the Toyota production system refers to this manufacturing technique.

It is of course important to consider the production technique in order to obtain the effect of changing costs by changing the method of manufacturing. But we must keep in mind that in today's world, the difference in production technique in whatever industry is insignificant. One element that can make a major difference is the manufacturing technique. By effectively utilizing equipment, personnel and raw materials, a substantial change in the cost can result.

When You Say "I Can't" You Admit Your Own Ignorance

We often meet a foreman with a neat white cap when we visit the workplace. He may have worked in the assembly line for thirty years, or he may have been in metal stamping for twenty-five years. People like him are living dictionaries for the workplace.

When machines or parts malfunction, the man in the white cap can discover right away what is wrong. Other workers may try to adjust them with unsteady hands, but the foreman comes with a hammer and strikes lightly, making the necessary adjustment.

Even in a process that requires high precision, the man in the white cap can adjust the machine to 1/1000-mm or 1/100-mm accuracy with ease. No one else can equal his expertise.

However, in spite of their great skills, these foremen tend to be unconcerned with the manner in which the work flows. "This line can plane 15,000 units," they will say, "and that has been our best record. Are you saying we must plane 17,000 units? No, we can't do it. Order 2,000 units from an outside source."

There are some metal-stamping mold makers with the same dilemma. Normally they create excellent molds. But once the amount is increased, they manufacture defective molds. Their schedule is disjointed and they do not know when they can deliver the extra molds ordered.

These are common occurrences. They do have excellent production techniques to produce molds, but they lack manufacturing techniques to let the entire work flow smoothly and to utilize effectively their equipment, personnel and raw materials.

Many people in the workplace will say: "We don't have the capabilities. We don't have enough people to do it." Change the manner in which things flow and change the manner in which you arrange your storage, and you will discover within a month that you can do what you have been saying you cannot do. Not only can you do it, but you will be left little extra change after paying the bill. In fact, you can even eliminate some of the processes!

> ## SAYINGS OF OHNO
>
> *A man-hour is something we can always count. But do not come to the conclusion that "we are short of people," or "we can't do it."*
>
> *Manpower is something that is beyond measurement. Capabilities can be extended indefinitely when everyone begins to think.*

To Work and To Move

To engage in a job means to work. In Japanese, the verb *hata raku* means to work. Someone has said that to work is to make people around you *(hata)* happy *(raku)*. At Toyota, we define the term *to work* very precisely. It means that we make an advance in the process, and enhance the added value.

Therefore, the term *to work* is used only when a certain action is definitely carrying forward a process or enhancing the added value. We do not call it work when someone is engaged in picking up something, putting down something, laying one thing on top of another or looking for something at the workplace. That is merely making a motion.

It is not that the Japanese people are especially diligent in their work habits, but they do feel uncomfortable when they have nothing to do at their place of work. After all, they are paid to do something, and for want of something productive to do, they engage in unnecessary motions. Thus, in work, there are two types of movement. One is the movement necessary for making products, one that moves the manufacturing process forward, and the other is not. The latter, of course, is a wasted motion.

Factories are equipped with chutes and conveyor belts to connect separate manufacturing processes. But what we often see in these factories is a site where workers place parts and materials two or three columns abreast on a chute or a conveyor. If there is only one item, a roller conveyor (or any other type of conveyor) can move with ease. But its movement is hindered when things are placed

Figure 1. To Work and To Move

side by side or are scattered along the conveyor. When the subsequent process tries to pick up the materials it needs, it has to engage in a lot of unnecessary motion to do so.

When the subsequent process picks up one item, other items may fall off the conveyor. Workers involved may be worried about their fingers being caught. All that tension and work are produced in picking up some items needed. It hardly seems worth the effort.

To pick up something or to replace something means simply that we change the location of certain items. We are merely moving them three centimeters away from the center of the earth or one meter closer to it!

What is important and what is not important, then? When we have this frame of mind, it becomes easier to differentiate the work load in the workplace. We may suddenly discover that only about one half of what we are doing is real work. We may give an appearance of working hard, but half our time is merely making moves without engaging in work. We move about a lot. This is a terrible waste and it must somehow be eliminated.

Reducing man-hours means to lower the waste and increase the amount of actual work. It does not mean that the size of the circle in Figure 1 has to be enlarged. And it is totally different from a movement to make the workers work harder.

> ## SAYINGS OF OHNO
>
> *Moving about quite a bit does not mean working. To work means to let the process move forward and to complete a job. In work there is very little waste and only high efficiency.*
>
> *Managers and foremen must endeavor to transform a mere motion* (ugoki) *into work* (hataraki).

Enhancing Labor Density

Generally people associate man-hour reduction with making the workers work harder. At Toyota, our thinking on labor density and making the workers work harder is as follows:

An example of making the workers work harder occurs when the work load is increased without improving the work process itself. For example, in a place which has been producing, ten units per hour, the company orders that henceforth fifteen units be produced, without improving the work process or equipment. If we try to illustrate this, it is like putting a bump on someone's head (or on a circle, as shown in Figure 2).

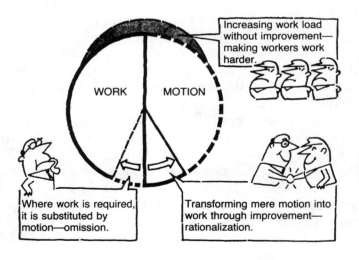

Figure 2. Enhancing Labor Density

In contrast, rationalization through man-hour reduction changes the wasted motion *(ugoki)* into work *(hataraki)* through improvement.

An act of omission *(tenuki)* occurs when someone does not do what he is supposed to do. For example, a plate must be secured tightly with five bolts, but a worker nonchalantly places four or five bolts without tightening each of them sufficiently. That is an act of omission or *tenuki*.

Toyota's man-hour reduction movement is aimed at reducing the overall number of man-hours by eliminating wasted motions and transforming them into work. All of us have some notion of what our work consists of. This movement eliminates from our work those actions which do not produce profit and which do not advance our process. It is a movement which channels the energy of men into effective and useful work. It is an expression of our respect for humanity.

Employees give their valuable energy and time to the company. If they are not given the opportunity to serve the company by working effectively, there can be no joy. For the company to deny that opportunity is to be against the principle of respect for humanity. People's sense of value cannot be satisfied unless they know they are doing something worthwhile.

At times, man-hour reduction has been considered to be merely an imposition of harder work without respect for humanity. This is due in part to misunderstanding and in part to the wrong method of implementation.

Based on what we have discussed so far, we can now define *labor density* as:

$$\frac{\text{Work}}{\text{Motion}} = \text{Labor Density}$$

The denominator is an impersonal motion and the numerator is work with a human touch. The act of intensifying labor density or of raising the labor utility factor means to make the denominator smaller (by eliminating waste) without making the numerator larger. Ideally labor density must be at 100 percent.

$$\frac{\text{Work}}{\text{Motion}} = 100\%$$

Around 1971, Toyota Motors' slogan was: "Eliminating waste to bring about improvement in efficiency." This was another expression of trying to make the denominator smaller.

Utility Factor and Efficiency

In manufacturing industries, elimination of waste is tied in with a better utility factor. As a result, if one can manufacture more products and parts than before, then one can say that the efficiency has increased.

Utility factor and *efficiency* are measuring sticks that we use daily. If we misuse these measuring sticks, we will be depriving ourselves of our ability to make the right evaluation. In fact, we can be faced with a situation in which efficiency has risen along with the cost.

The term *utility factor* is defined as the percentage of the energy supplied to a machine relative to that machine's actual capabilities. It can never be expressed in a number larger than 100 percent. When this definition is applied to production, the utility factor in production becomes the percentage of labor expended for producing a given product in relation to the labor required for making that product.

When the utility factor in production is 50 percent, it means that only one half of the worker effort is useful in making that particular product. The remaining 50 percent is being wasted. When the utility factor in production is 80 percent, it means that 80 percent of the worker effort is useful, and the utility factor in production is much higher than in the previous example.

Thus any production which has a high utility factor means that most of the labor expended is going into the power to produce a given product.

In contrast, the term *efficiency* is used when one wishes to compare output. That is, within a given time frame, how many people have produced how many pieces? To compare, one needs a set standard (criterion). Normally, the actual performance of the past, such as the past month or year, is the standard. Or a company may set an arbitrary standard and say: "This month we have raised our efficiency by 15 percent (relative to our standard)." Thus, unlike the utility factor, efficiency may exceed 100 percent.

Don't Be Misled by Apparent Efficiency

At one production line, 10 workers made 100 pieces every day. As a result of improvement, the daily output has increased by 20, to 120 pieces. This has been called a 20-percent improvement in efficiency. But is it?

When efficiency is expressed in an equation, it becomes:

$$\text{Efficiency} = \frac{\text{Output}}{\text{Number of workers}}$$

Generally, when there is talk of raising efficiency, most people think in terms of increasing the output (the numerator in this equation).

It is relatively easy to increase the number of machines or the number of workers in order to increase the output. Also, by sheer determination, all the workers can work together to raise the output. In a period of high economic growth, or in a company that is experiencing an increase in sales, either one of these approaches would, of course, be fine. But in another period and in a different company, can either one of these approaches work?

In a recession, or when the company's sales are declining, can it continue to allow this particular line to produce 100 pieces daily as part of the company's production plan? Can the line insist on producing 120 pieces because of its efficiency, even though the production plan calls for reducing the output to 90 pieces? What can the company do with the 20 to 30 pieces that are overproduced daily? The overproduction forces the company to pay for the costs of unnecessary raw materials and labor. Then there are costs of pallets (for storage and transporting) and of storage areas. For the company, overproduction means a net loss. Improvement in efficiency which does not contribute to the company's overall performance is not an improvement but a change for the worse.

Now, using the same example, but where the output needed does not change or is reduced, how can a company secure an improvement in efficiency that will assure profitability?

In such an instance, the process must be changed in such a way that only 8 instead of 10 workers will be required to produce 100 pieces. (Or if the quantity required is 90 pieces, let 7 workers handle the task.) In this way, efficiency will improve and be accompanied by a reduction in costs.

When we speak of attaining a 20-percent improvement in efficiency, there are two ways of doing it. It is easy to increase the number of machines to raise efficiency. But it is several times more difficult to reduce the number of workers and still raise efficiency. No matter how difficult the latter may be, we must take it up as a challenge. This is especially important in a period of recession, when we must attain efficiency through man-hour reduction.

Toyota does not allow an increase in output to create an appearance of efficiency improvement, when there is a need to reduce production or maintain the same output. We call it *an efficiency improvement for the sake of appearance*.

SAYINGS OF OHNO

When output needed does not change or must be reduced, do not attempt to improve your efficiency by producing more. Do not engage in an efficiency improvement for the sake of appearance.

No matter how difficult it may be, take it up as a challenge to reduce man-hours as a means of improving efficiency.

It's a Crime to Overproduce

What the Toyota production system seeks is a total elimination of waste.

We say that "a manufacturer's profit can be found in the way he makes things." It reflects our philosophy of attaining a cost reduction through the elimination of wasteful operations. There are many types of wastes. At Toyota, in order to proceed with our man-hour reduction activites, we divide wastes into the following seven categories:

1. Waste arising from overproducing
2. Waste arising from time on hand (waiting)
3. Waste arising from transporting
4. Waste arising from processing itself

5. Waste arising from unnecessary stock on hand
6. Waste arising from unnecessary motion
7. Waste arising from producing defective goods

The most common sight found in many workplaces is the excessive progression of work. Everything moves too fast. Normally, it must be consigned to waiting, but workers proceed to the next stage of work. Thus the time that is supposed to be waiting time becomes hidden. When this process is repeated, materials or parts produced accumulate in between or at the end of the production line, creating unnecessary stock on hand. To transport this stock or to rearrange it for storage requires creation of another type of work. By the time this process takes its course, it becomes more and more difficult to find where the wastes are.

Under the Toyota production system, we call this phenomenon the *waste arising from overproducing*. Of the many infractions of wastefulness, this is considered by far the worst offense.

The waste arising from overproducing is different from other wastes, because unlike other wastes, it overshadows all others. Other wastes give us clues as to how to correct them. But the waste arising from overproducing provides a blanket cover and prevents us from making corrections and improvements.

Thus, the first step in any man-hour reduction activity is to eliminate the waste arising from overproducing. To do so, production lines must be reorganized, rules must be established to prevent overproduction, and restraints against overproduction must become a built-in feature of any equipment within the workplace.

Once these steps are taken, the flow of things will return to normal. The line will produce one item at a time as needed. The waste becomes clearly discernible as the waste arising from time on hand. When a production line is reorganized in this fashion, it becomes much easier to engage in the activity consisting of "elimination of waste — reassignment of work — reduction of personnel."

The waste arising from time on hand (waiting) is created when a worker stands idly by an automated machine to serve as a watchman, or when he cannot do anything constructive manually because the machine is running.

This waste is also created when the preceding process fails to deliver parts needed in the present process, thus preventing workers in the latter from working.

In the illustration below, a worker is assigned to each of the machines designated as *a, b* and *c*. In this process, the worker stands by idly while the machine moves. He cannot work, even if he wants to, and there is a waste arising from waiting.

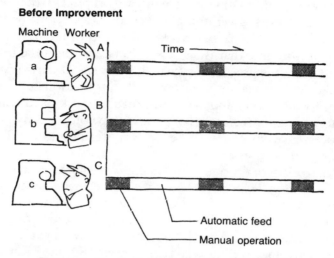

Before Improvement

Machine Worker

Figure 3. Waste Arising from Time on Hand

In order to eliminate this waste, the worker named A is assigned to all three of these machines to operate their automatic feeds sequentially. Under this arrangement, worker A places material in machine

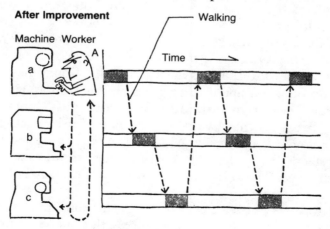

After Improvement

Machine Worker

Figure 4. Eliminating Waste Arising from Time on Hand

a and pushes the starter switch and moves to machine *b*. He places material in machine *b* and starts it. He moves on to machine *c*, and repeats the same process he has done with machines *a* and *b*. After *c* is started, he moves back to *a*. By the time worker A returns to machine *a*, the work there is completed and he can immediately start another round of work on machine *a*.

By eliminating the waste arising from waiting, two workers can be removed from the work process. Similarly, one may also consider eliminating unnecessary motions, which do not contribute to the work itself.

The waste arising from transporting refers to waste caused by an item being moved a distance unnecessarily, being stored temporarily or being rearranged. For example, traditionally parts are transferred from a large storage pallet to a smaller one and then placed temporarily on a machine several times before they are finally processed. By improving the pallets, we have been able to dispense with these temporary placement procedures and let one worker operate two machines.

Another instance of waste arising from transporting occurs when parts are moved from a warehouse to the factory, from the factory to the machines and from the machines to the hands of workers. At each of these steps, parts have to be rearranged and moved.

The waste arising from processing itself occurs, for example, when a guide pin in the jig does not function properly and the worker

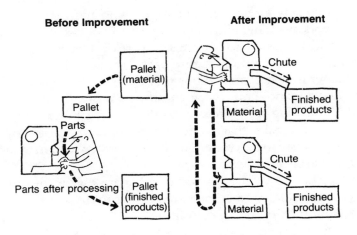

Figure 5. Eliminating Waste Arising from Transporting

has to hold the jig with his left hand. The processing does not go smoothly and time is wasted.

In addition, there are wastes arising from unnecessary stock on hand, from unnecessary motion and from producing defectives. Explanations for each of these are not necessary.

SAYINGS OF OHNO

A worker or a line with excess capacities inevitably moves forward if left alone. When this happens, wastes are hidden.

In other words, overproduction creates a countless number of wastes, such as over-staffing, pre-emptive use of materials and energy costs, advance payment to workers, interest charges on mechanical devices and products, storage areas needed to accommodate the excess products and the cost of transporting them.

In a period of low economic growth, overproduction is a crime.

Eliminate Waste Thoroughly

There are many foremen and managers who allow their subordinates to work on something they suspect to be wasteful. Many of them consider such an act a necessary part of their job, and often they do not understand the nature of the waste.

No matter how determined one may be in his desire to eliminate waste, if he does not know what constitutes waste, then there is no way of eliminating it. Therefore, an important task awaiting each of us is to make sure that waste always appears — distinctly and clearly — as waste to everyone. This is the first step toward attaining an improvement in efficiency.

Among the many types of waste, some are easy to discern and others are difficult. Among them, the easiest to discern is the waste arising from time on hand, or waiting.

For example, if the cycle time is three minutes and there is a one-minute period of waiting before a worker can resume his work, the worker himself, his supervisor and other superiors certainly will know that this one minute is wasted. However, if the worker moves around to spend this one minute as if he were working, no clear-cut image will emerge (the wastes of transporting and of processing itself). Or if he should use this time to process the next item, no one can tell if a waste has actually occurred (the waste of overproducing). All these three wastes must be translated back into the waste arising from waiting. It can facilitate devising appropriate countermeasures.

In this connection, we may consider taking the following three steps:

1. Let the workers strictly observe the standard operations. Do not allow any deviation.
2. Control excessive forward movements through the production system, which utilizes kanban.*
3. Clearly indicate on the conveyor line the work area for each worker, thus preventing any worker from moving ahead of the schedule to do excessive work.

We shall explain these steps in succeeding chapters. An important thing to remember is that to eliminate waste, you must find it first. You must rearrange the workplace in such a way that waste can be easily found. Each of the steps you take by themselves seem insignificant. For example, you may even have to worry about a small quantity of goods stored between two processes. But as long as the problem is related to your "efficiency enhancement – cost reduction" movement, you must be prepared to ask this question: "Why has this occurred?" Before long, you may find a clue to the very improvement you are seeking.

Enhancement of efficiency can be attained through elimination of waste. Of course, there are many ways of finding different wastes. But the most effective way remains that of *translating such wastes into the waste arising from waiting*. It is one that is easy to detect, and provides the first step toward efficiency enhancement.

This total dedication to the elimination of waste is the heart and soul of the Toyota production system. It also constitutes the very source of its profit.

---

* A *kanban* is a signboard or card; the word also refers to the system utilizing standard containers, each of which has a card designating what and when to produce. See Chapter 5.

2
Basic Assumptions Behind the Toyota Production System

Toyota Production System and Kanban System

Many people may immediately associate the Toyota production system with the kanban system. While this is not wrong, it is not exactly accurate.

The *kanban system* is one of the methods of control utilized within the Toyota production system (the way we make things). One cannot discuss the kanban system out of context. If anyone tries to imitate that system without regard to all the factors contributing to its success, then his efforts will be in vain.

The Toyota production system is unique and unparalleled. The thinking behind it and the method of implementation have been perfected after long years of trial and error.

In a nutshell, it is a system of production, based on the philosophy of total elimination of waste, that seeks the utmost in rationality in the way we make things. We call this the Toyota-style production system or the Toyota production system. Hereafter in this book, we shall use the term *Toyota system* to represent it.

Only when the Toyota production system in its totality is satisfactorily conducted, can there be an effective utilization of the kanban system. Without changing the method of making things, it is impossible to engage in the kanban system.

Please take note of this fact before proceeding further in this chapter.

An Outline of the Toyota System

We have prepared a chart to provide a bird's-eye view of the Toyota system. It is reproduced on the following page.

An ideal condition for manufacturing is where there is no waste in machines, equipment and personnel, and where they can work together to raise the added value to produce profit. The most important concern for us is how closely we can approach this ideal.

To make the flow of things as close as possible to this ideal condition — whether they be between operations, between lines, between processes or between factories — we have devised a system in which the materials needed are obtained *just-in-time* — that is, exactly when needed and in the quantity needed.

On the other hand, for this ideal condition to occur in the line operations, including machines and equipment, if there is abnormality, everything must be stopped immediately at the discretion of the worker or workers involved. (Machines must be endowed with the same faculty.) The reasons for the occurrence of abnormality must be investigated from the ground up. This is what we call *automation with a human touch*.

We believe it is best to manufacture everything in a balanced manner. This *load-smoothing production* serves as the base for the two pillars of the Toyota system, namely the just-in-time and automation-with-a-human-touch approaches.

Characteristics of the Toyota System

Now that we have a general notion of the structure of the Toyota system, we may proceed to enumerating the characteristics of this system. In this way, we shall be able to discern the basic ideas behind the Toyota system.

A Company-Wide IE Activity Directly Connected with Management

There is no specific formula for a manufacturing method that can apply to all products in all processes. Therefore a product may be manufactured by one worker in one company while at another it may take two workers to produce the same. In a company that

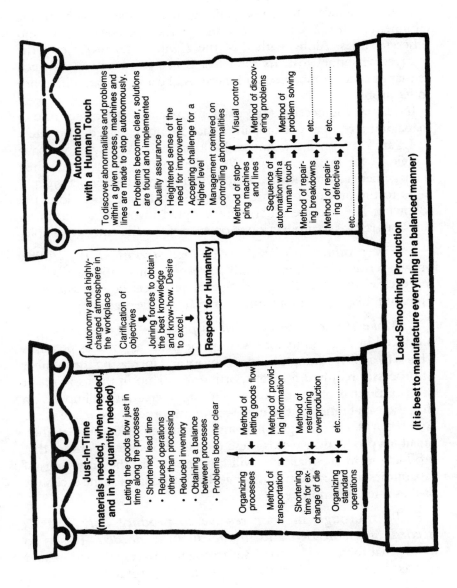

Figure 6. The Two Pillars of the Toyota System

is not concerned with the method of production, it may even require three workers to do the same work.

In this instance, the company using three workers must assume a higher cost for its warehouse, transporters, pallets, conveyors and other facilities. Along with this, there will be an increase in the indirect personnel cost. Its total cost is likely to be twice as much as that of other companies, and there will be a significant difference in its profit.

Industrial engineering (IE) plays an extremely important role in managing a company. Companies that are not engaged adequately in IE are very much like houses built on sand. At Toyota we have a saying: "IE makes money." We assign to IE a firmly established management role to enhance improvement in production activities.

As to the production system, our basic thinking goes along the following lines. These criteria are adopted in order to permit all divisions related to production to operate more efficiently as a whole.

1. *The production plan must be load smoothed.* If we think only in terms of the final assembly process, it may appear that it is more efficient to let the same type of products flow at the same time. But this will create a number of wastes in the preceding process.

2. *Make the lot size as small as possible.* Metal stamping is produced by lot, and its lot size must be made as small as possible. This is to avoid creating a large inventory and increasing the number of processes required for transporting. A mixup in the assignment of priorities has often resulted in shortages, creating an impression that the capabilities of the metal-stamping process are inadequate. As a result, some managers may insist on installing additional lines. Small-lot production avoids these pitfalls. However, to ensure that production in a small lot will not result in the lowering of capabilities, it is recommended that the procedures for the exchange of dies must be improved.

3. *Be thorough in your resolve to produce only what is needed, when needed and in the quantity needed.* This is to ensure that there will be no waste arising from overproduction, and to make it clear to everyone that the process has an excess capacity.

Scientific Attitude Emphasizing Facts

At the workplace, we start from the actual phenomenon, investigate the cause and find a solution. We do not deviate from this approach. In other words, anything related to the workplace is based on facts. No matter how much information is provided through data, it is difficult to see the true picture of the workplace through data. When defects are produced, and we find out only through data, we miss the chance to take appropriate corrective actions. Thus we may not be able to discover the true cause of the defects, resulting in our inability to take effective countermeasures against recurrence. The place where we can accurately capture the true state of the workplace is the workplace itself. We can catch defects on the spot in the workplace and then find the true cause. We can immediately take countermeasures. That is why under the Toyota system, we say that data is important but we emphasize facts even more at the workplace.

When a problem occurs, if the manner of probing into the cause is insufficient, measures taken can become blurry. At Toyota, we have the so-called five W's and one H. The five W's are not the conventional "who, when, where, what and why," but every word is replaced by a "why," and we say "why, why, why, why, and why" five times before we finally say "how?" In this way, we delve into the true cause that is hidden behind the various causes. It is essential that we come face to face with the true cause.

To make this method thoroughly understood by all, we take the following steps:

1. *Make sure that everyone can understand where the problem is.* If we know where the problem is, it is relatively easy to solve. Quite often difficulty arises because we cannot determine what the problem is. Thus we frequently use the *kanban* and *andon*. (The word *kanban* refers to the signboard of a store or shop, but at Toyota it simply means any small sign displayed in front of a worker. The word *andon* refers to a Japanese paper-covered lamp stand, but at Toyota it simply means a lamp. Hereafter the word *kanban* will be used throughout. The word *andon* may either be used in its original form or translated as a display lamp, as the occasion demands.)

2. *Clarify the purpose behind the task of problem solving.* We probe into the true cause and offer a solution. If we do not probe deeply into the true cause, we may be offering merely a temporary solution, which cannot result in prevention of recurrence.

3. *Even if there is only one defective item, provide a corrective measure.* Even if the defect occurs only once every thousand times, ascertain the facts. With these facts, the true cause can be found and steps can be taken to prevent recurrence of defects. This type of defect is harder to find than those that occur more frequently. Be attentive, and do not overlook it when it occurs.

The Man-Hour Reduction Activity Must Be a Practical One

A step-by-step approach is required in this. The goal may be set high, but its implementation calls for progression in stages. We also place a great deal of emphasis on results. From these two basic assumptions, the following considerations emerge:

1. *Move from work improvement to equipment improvement.* Toyota insists on carrying out the work improvement phase of improvement activities thoroughly before moving on to the equipment improvement phase.

When good results can be obtained with work improvement, and when such work improvement has not been undertaken sufficiently, there is no justification for investing a huge sum in automation machines. The effect of the introduction of automation machines may roughly equal that of a thoroughly conducted work improvement activity. In such a case, the money spent for equipment investment is wasted.

2. *Differentiate between man-hours and number of workers, and between labor saving and people saving.* In calculating the number of man-hours required, it is possible to say that a certain process requires 0.1 to 0.5 worker. But in reality, the work requiring only 0.1 worker still needs one person. Thus if the work load of one worker is reduced by 0.9 worker, it still does not result in any cost reduction. True cost reduction can come about only after the number of workers is reduced.

Therefore, when we engage in a man-hour improvement activity, we must focus our attention on reducing the number of workers.

When automation devices are installed, there may be a labor saving of 0.9 worker. But if the process still requires 0.1 worker, the money spent does not result in reducing the number of workers. This is often erroneously looked upon as labor saving. In order to avoid confusion resulting from the use of this term, Toyota refers to a reduction in the number of workers, which can truly bring about cost reduction, as *people saving* to differentiate it from *labor saving*.

3. *To check means to give thought to something.* An improvement activity is completed when the result sought in the initial goal is obtained.

If the result cannot be obtained, it is often that the work has been done without much thought given to it. Confirm the result of implementation at the workplace, adjust all those parts inadequately done and confirm the result again. By repeating this process, a good result can be obtained through improvement.

When we check things, we do not merely look them over. It must be a process through which we rethink and reflect on our own work.

"Economy" Is Everything in the Standard of Judgment

The objective of the man-hour reduction activity is to lower the cost. Therefore, in every thought process must be a yardstick that asks: "Which one is more economical?" In its practical application, there are these considerations:

1. *The ratio of operation for equipment is determined by the quantity required for production.* Some people say that the higher the ratio of operation, the better, and they may overproduce those items that are not needed every day. They must store the excess products, and the loss resulting from overproduction is far greater than if they had produced only what was needed. It is dangerous to establish a standard based on raising the ratio of operation. Do not ignore the fact that the ratio of operation of machines and equipment must be based on the quantity required for production.

2. *When you have time on hand, use that to practice exchange of die.* The wages of workers who have established working hours but have nothing to do remain the same whether they remain idle or engage in the practice of exchange of die. If there is any time

left, utilize that time to practice exchange of die, which is relatively complicated, or to train a not-so-skilled worker to become skilled in the standard operations in that particular area.

The Workplace Is the Boss

We consider the workplace to be an organic entity. The hands and legs have not entrusted their brains to the management division. Therefore, engineering must not act as if it were the commander-in-chief of the workplace. On the contrary, the autonomy of the workplace must be emphasized and respected. The engineering division gives support to the workplace and provides services in those areas that need them. It makes certain that the responsibility is not scattered and that the information provided is neither excessive nor deficient.

Emphasize Immediate Response to Change

Once a plan is established, it is often forced to change due to external and internal conditions. If the workplace insists on carrying out the original plan, distortion will occur, and it might adversely affect the rest of the company.

The workplace must establish a system that can respond quickly to changes forced upon it through the interplay of external and internal conditions. The greater the ability to respond to change, the stronger the workplace.

For example, due to an increase or decrease in production or to the stopping of the production line, the plan previously given to the workplace has to be changed. If the workplace can immediately establish the best system possible to solve all the problems quickly, and in the process show no confusion, then it is showing the ideal style of operation. We call such a place a workplace with its own soul.

The Goal Is Cost Reduction

The Toyota system is a series of activities that promote cost reduction through the elimination of waste to achieve enhanced productivity. All company-wide improvement activities must directly contribute to the goal of cost reduction.

The various methods for improvement and the thoughts behind them, in the final analysis, must be related to cost reduction. Conversely, cost reduction becomes the basic criterion on which we base our judgment.

If this basic criterion is not clearly understood, some managers may become unthinking advocates of improvement. Overanxious about improvement after improvement, they may end up creating the waste arising from overproduction.

A company may spend money to improve its equipment and machines and spend time to improve its operations, yet find, in the end, that its only increase is excess inventory. The more they do, the worse off they become. That is an act of improvement contributing to a company's demise.

It is easy to say "cost reduction," but in making a decision there are two avenues, and they must be clearly differentiated. The first is a question of judgment, determining which is more advantageous between A and B. The second is a question of selection, choosing the most economically advantageous plan from among plans A, B, C and others.

One Goal, Many Approaches

Should a product be manufactured by the company or subcontracted? Should a company purchase a machine for the exclusive use of a certain process, or should it concurrently utilize for this particular process a machine it presently owns and uses for other purposes? These are questions of judgment, and the company must decide whether A or B has greater overall merit for the company.

Now, let us consider the question of selection, choosing a plan for its economic advantage among the many options.

For example, the goal is to reduce manpower, for which there are many approaches. Automation can reduce manpower, and so can restructuring the work process. Or a robot may be introduced. The company must study carefully all of these options to determine which is most advantageous.

Let us assume that there is a plan which suggests that an electric control device, costing $500, be installed to reduce the number of workers by one. If this is implemented, with a mere $500 Toyota will be able to reduce its workers by one. This represents a considerable saving for the company, and sounds like a good idea. But is the device really necessary? On closer examination, we see that by

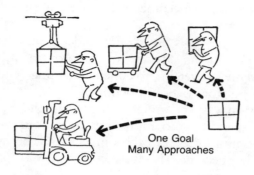

One Goal
Many Approaches

Figure 7. One Goal, Many Approaches

changing the work sequence, one person can still be removed from that particular process. The $500 is therefore actually wasted. It is a premature plan that is a failure. We cannot select a plan simply because it can save some money. We must select a plan that can save more than other plans can save. Often, without thinking, companies choose automation. Beware of the pitfall just mentioned.

In undertaking the task for improvement, as discussed earlier, during the investigation stage there are two alternatives. The important thing to remember is that there are many approaches and methods to reach the same goal. So, carefully study as many plans as possible, take into account the company's overall objectives, and then select a plan best suited for the particular process.

Do not proceed with your improvement activity without full investigation. It can turn into an improvement plan that costs too much. Be sure to keep this fact in mind always.

SAYINGS OF OHNO

Every decision must be based on these principles: "Can the cost really be lowered?" and "Can this action help the company's overall performance?"

Excess Capacity and Economic Advantage

Making a decision on economic advantage can be swayed by the existence or non-existence of excess productive capacity. If there is excess capacity, the company merely uses workers and machines that are not occupied. There is no new expenditure, and the entire process can be free to the company.

Producing within the company or subcontracting: Should a certain part be produced within the company or subcontracted? Often management compares the relative costs. However, if there is excess capacity within the company, the new cost created is merely a fluctuation in the cost of materials and energy. No comparison of costs is necessary in this case. Production within the company is to its advantage.

Using stock on hand: A worker who transports parts from one line to another is waiting until the pallet is filled. To let him engage in the line work or preparatory work does not raise the cost. There is no need to study profit and loss in this situation. Do not add the man-hours spent here as an increase in man-hours.

As these examples demonstrate, when there is excess capacity, cost accounting is not needed to show which action is more advantageous to the company. It is important to make the existence of excess capacity known at all times. If this is not made clear, managers are liable to make a wrong decision and raise the cost for the company.

What Is Effective Utilization?

Facilities and personnel are more than adequate but they remain idle because there is no work. This is a fairly common sight in many companies.

When this type of situation develops, the reaction is often one of "well, we can't do anything about the machines, but it's wasteful to let the workers remain idle." So managers may order the workers to mow the lawn or clean the windows. This is the wrong approach.

These managers may have effective utilization of idle workers in mind. But no matter how meticulously the lawn is mowed or the windows kept clean, it does not produce a dollar of profit. Effective utilization must at least contribute to cost reduction. This is especially important when no work can be found for the workers and there is no way to increase the added value for the company.

At a certain factory, there was no longer any work available and workers were left with literally nothing to do. It so happened that at a number of locations there were water leak problems that were left unattended when the factory was in full operation. The company decided to fix the leaks during this period. The following month and thereafter, the water bill was trimmed by $5,000 each month. This is an example of true effective utilization.

Is It a Waste, If You Do Not Use an Expensive Machine?

Many people have an erroneous notion that an expensive piece of equipment already purchased must be fully operated to get their money's worth. The higher the purchase cost, the greater its depreciation. So there is a feeling that unless the ratio of operation, or machine utilization, is close to 100 percent, money is lost.

However, while it may be true that the higher the ratio of operation the better, the loss resulting from overproduction may become far greater if the factory produces something that is not needed. Therefore, as stated earlier, it is dangerous to set the criterion solely on the basis of raising the ratio of operation. One must not ignore the fact that the ratio of operation of machines and equipment must be based on the required amount of production.

At Toyota, we are thoroughly committed to the idea that we must respect the work of our people at the core, usually hidden behind the many machines. In other words, we are people-centered and not machine-centered. If we become machine-centered, we may overproduce and create an excess of workers. If we develop our work schedule in a people-centered manner, we will be able to adjust the ratio of machine operation and eliminate the waste arising from excess workers. We can achieve this by making our work consistent with the required output or demand, and by operating the machines accordingly.

The money already spent is called *embedded expense,* or *sunk cost,* and cannot be used for future plans. When one is thinking of improvement, do not consider this to be a restraining factor. Many mistakes have arisen because of it.

For example, there may be a feeling that money is lost if a high-priced or high-performance machine is not used. But in principle, as long as a machine is in the workplace, irrespective of its high or low cost, the usage that the workplace has of the machine and its price are not related. If there is an issue arising whether to use a high-priced or a low-priced machine, simply use the one that costs less to operate.

High Speed and High Performance Can Be a No-No

Car seats are sewn together by industrial sewing machines. Some lines are straight and some are curved. When one stands by the side of such a machine, he hears the sounds: JA-JA-JA-JA-JA, JA-JA, JA-JA-JA-JA-JA, JA-JA. . . . The sound of the sewing machine changes and is interrupted as the worker sews the straight line, curved line or a more complex patch.

Sewing machines used to be foot-pedaled. Today the machine is powered by a motor and its speed is fast. It uses a clutch to connect and disconnect power, just as in a car.

Most unskilled workers will not have any problem in guiding a piece of cloth forward when it is sewn straight, and they usually sew that part in one stretch. For curved lines, they cannot move the cloth forward consistently with the speed of the machine. So they slow the speed and their work is represented by the interrupted sounds of JA,JA,JA.

For skilled veteran workers, their rhythm is about the same whether the lines are straight or curved. On the other hand, they do not sew the straight line part in one stretch. Their motion is actually slightly slower than that of the inexperienced workers, but with an even JA-JA-JA-JA sound. With the same pace, they then proceed to sew the curved line part.

This is so because the more experienced workers become, the better is their clutch action. In a sense, they control the machine's mechanical speed with their personal touch, dropping the machine's speed to meet the requirements of their work.

Traditionally, industrial machines had to sew thick and hard materials, and their speeds were not that fast. But thanks to new technology, their speeds have become much faster, and today they have literally become machines of high speed and high performance.

However, along with their speed and performance, the price of these machines has also risen.

Yet these expensive machines may stop frequently in the hands of inexperienced workers. And the veterans lower their speed in order to meet their work requirements. Why is it necessary to buy these expensive machines in the first place?

Taking this into account, Toyota asked its cooperating company to manufacture lower speed machines. The cost of these was half that of the higher speed machines.

A Little Over Time Adds Up

The cost that recurs every day looks inconspicuous enough every time one sees it, and it is easy to overlook. Conversely, the one-time-only cost often appears to be a very expensive one because the sum for one time is rather substantial. However, if we convert into a one-time expenditure the cost caused by the waste which continuously occurs over a two-year period, we will be unpleasantly surprised by how large a sum it really is. We may feel that the one-time expenditure needed for improvement is too high (aside from the desire to find less expensive alternatives), but if we neglect elimination of waste that continuously occurs, we may incur a greater loss in the end. We cannot be led by a vague feeling. Instead we must count. In setting up lamps and kanban, there will always be an added cost. Or when the improvement is implemented, on a strictly temporary basis, overtime work may increase. Debates over these issues always surface, but we must always remember to count carefully and accurately.

How to Use Yardsticks

Some familiar sayings, such as "The x ratio has increased," or "This method yields a higher y ratio and is therefore more advantageous to us," may sound fine. But depending on the goal the company seeks, this type of thinking may actually cause poor judgment. Yardsticks such as the profit ratio of a product or a particular type of investment are useful tools. Yet there are times when these yardsticks cannot be used to select a production or investment plan advantageous to the company.

I have already referred to the ratio of operation a number of times. It is wrong to assume that a decline in the ratio of operation equals a loss. The most profitable way, and at the same time the least wasteful way, is to manufacture products at the time needed in the quantity needed. If a company is too concerned with the ratio of operation and insists on operating all machines at 100 percent capacity, it may be left with surplus finished and semi-finished products piling up here and there. It will probably require two to three times the manpower just to handle these excess products. The company will also be forced to buy several times more materials and parts than required.

When this is seen strictly from the perspectives of expenditure and income, the amount expended is likely to register three to four times above normal, while the amount of income remains constant. In such a case, the term *loss* will sound too moderate.

Therefore it is best to consider that the ratio of operation is determined by the required output. However, the factory must always be ready to operate whenever called to do so. Otherwise, it will result in a loss from lost opportunities — or a necessity for overtime, which is also a loss.

There are several ways of expressing in a ratio the relationship between the results of work and labor. In addition to the commonly used terms such as *utility factor* and *efficiency,* there are phrases such as the *ratio of operation, labor productivity* and *strokes per hour (SPH).* All of these are yardsticks designed to evaluate how efficiently the work is progressing.

In using these yardsticks to measure the results of work, consider the following:

1. *To raise the ratio of operation or SPH in itself cannot be the goal of the company.* Our goal is to reduce cost. Raising the ratio of operation or SPH without regard to all existing conditions can often result in a higher cost. For example, a line can raise its ratio of operation somewhat by any of the following methods: by allowing each process to have semi-finished products that can cover for equipment failures; by holding all types of parts in abundance, so as not to be affected by the shortage in the preceding process; and by assembling those goods for which all parts are on hand. But the past thirty years of experience in workplace management have

shown us that these approaches often raise the cost. Therefore these approaches can be used only when they are consistent with the overall goal of the company. They can be used as yardsticks only when those employing them have a clear picture of all conditions present.

2. *The way you look at "capacity" is very important.* If you are speaking of machines and equipment, the highest capacity for them is generally represented by their *machine cycle* (or the period of continuous stamping). To evaluate properly the machine presently in use, ascertain its present capacity, and look at it with a view to how far it can be raised when needed. If you are talking about manpower, you must differentiate between mere movement and working. Do not consider it to be a worker's capacity, when he moves about with wasted motions.

3. *It is important to think about the concept of time that "faster" expresses.* To work faster can become meaningful only if more processes and fewer people can do the same job.

To produce faster means that more products are manufactured within a given time frame. Yes, the efficiency is enhanced. But at times it can also mean a loss for the company.

High Efficiency Does Not Equal Lower Cost

As discussed earlier, the purpose of enhancing efficiency is to lower the cost. Therefore enhancement of efficiency cannot in itself be considered a goal to be pursued. Only when higher efficiency and lower cost become one can there be meaning to the act of enhancing efficiency.

We often see production lines adopting as their management objective the enhancement of SPH (strokes per hour or per hour productivity). They place behind the line a production control board, which records the amount of finished products every hour on the hour.

When this is continued, one tends to confuse the enhancing of SPH with the objective itself.

To enhance SPH, a foreman or manager may decide to engage in production by a large lot, reducing the number of times the line engages in the exchange of die. After the day's quota is filled, if there is time left, the line may start producing the next day's or even the day after tomorrow's quotas. Indeed, SPH will rise and people who are directly involved may feel that their efficiency is high and they have made money for the company. But in reality what they have created is a mountain of stockpiled materials and parts between themselves and the subsequent process.

In this instance, the first condition for this line to observe is to produce only the amount needed, in as small a lot as possible. If they try to raise SPH within this framework, they can succeed in cost reduction.

Without abiding by this condition, if the line attempts to raise SPH, it only creates a net negative situation for the factory. High efficiency does not always equal low cost.

Ratio of Operation and Ratio of Movability

The *ratio of operation* refers to the ratio of how many hours during a working day a machine is utilized to manufacture. Since the work day is generally defined as eight hours, if a machine is operated only four hours, the ratio of operation of that machine is a mere 50 percent.

In Japanese we use three Chinese characters *(kanji)*, *ka-do-ritsu*, to represent the term *ratio of operation*. The character *ka* stands for the word *kasegu*, which means to make a profit, and the character *do* means to move. Therefore when we use the term *ratio of operation*, we literally expect the machine to be operated in order to make a profit. If a machine moves continuously for the entire day without producing anything, the ratio of operation will remain zero. As discussed earlier, we must differentiate very clearly between an act of "moving" and an act of "working." This concept is equally applicable to a machine. If there is wasted motion in a machine, or if hours are spent without producing real work, such waste must clearly be avoided.

Because of this, at Toyota we write the term *ratio of operation* with one extra radical added to the character *do* to signify real work with a human touch.

With all of these in mind, if we are to provide a new definition for the Toyota-style ratio of operation, we mean "the ratio of actual production to the capacity of a machine when it is fully utilized."

In other words, if machine A has the capacity for producing 100 pieces per hour of a given part, and if in a particular day only 50 pieces per hour are produced, then the ratio of operation for that day is 50 percent.

The ratio of operation fluctuates from month to month, influenced by the sales figures and the number of cars produced. If sales are down, so is the ratio of operation. Conversely, when orders are up, longer overtime hours and extra shifts may be required. If the normal full eight-hour operation is defined as 100, then the ratio must be raised to 120 or 130 percent.

This is why we say no factory can establish a certain percent ratio of operation as its production goal.

Toyota factories are very much like those of any other car manufacturer in that many machines stand side by side. But what distinguishes Toyota factories from others is that while some of the machines are operating, others are stopped.

Visitors are often heard to say: "How can you make money while keeping this many machines idle?"

Our response is simple. The way in which we organize our work is based on this principle: "Timing is the key to everything we do."

For example, there is a machine which is capable of cutting a piece in 10 seconds. If the machine is forced to cut each piece at a 10-second interval without interruption day and night, it may break down within a year or two. But if we time it so that a piece is cut every four minutes, the actual cutting time remains 10 seconds, but the machine is stopped for the remaining three minutes 50 seconds.

The term *ratio of operation* shares the same sound in Japanese, *ka-do-ritsu*, with another term, *ratio of movability* or *reliability*.

The ratio of movability represents the state in which a given machine functions properly when operated. When the switch is on, the motor turns, the machine moves and the operation proceeds. This is the normal state of the machine's operation.

Ideally, the ratio of movability must be 100 percent, and that must be set as the target.

In order to attain this, preventive maintenance must be done to avoid a breakdown. It is also necessary to shorten the time interval between the exchange of die.

Let us use your car as an example. The ratio of movability is whenever you want to take a ride, the car starts immediately, the engine moves smoothly and you can have a nice drive.

For a pleasure ride on a Sunday afternoon, the ratio of movability may not be a serious consideration. But suppose your child is suddenly taken ill, and you have to go to a doctor. The engine does not start, a tire is flat and there is no gasoline in the tank. There is no end to all the trouble you may have. That is why the ratio of movability must always be kept at 100 percent.

On the other hand, the ratio of operation is the number of hours you operate your car. You have finally bought the car of your dreams. But will you ride in it nonstop day and night? On weekends, you may take your family for a ride. But on an average day, you may take your wife shopping and operate it for an hour or two. And, unless you commute by car, you may not operate it at all on some days.

People drive their cars when needed. Thus, the 100 percent ratio of operation in this case is nonsense. Driving a car unnecessarily means a net loss. Aside from the expense for gas and oil, one must also consider the wear and tear on the car, which can result in an earlier breakdown.

SAYINGS OF OHNO

The ratio of operation is the burden imposed on a machine in relation to its capability when it is fully operated, and is determined by the sales figure. The ratio of movability is the condition of that machine to operate on demand. In this case, the ideal state is 100 percent.

Shorten the Lead Time

Whatever the process, be it the way in which machines are aligned or the manner in which materials flow, the longer the lead time for production, the worse it is for the process.

At Toyota, we define *lead time* as the time elapsed from the time we start processing materials into products to the time we receive payment for them.

For example, a certain product is said to take a month to produce. But if we take a closer look at it, we discover that the actual time the production processes take is extremely short. The time spent for manufacturing is far shorter than the time the product lies idle in storage.

Generally speaking, lead time is the sum of the time required for processing and the time the product is in storage. Thus in some quarters, fear is expressed that the ratio between the time required for processing and the time the product is in storage may become as high as 1 : 100.

When the lead time becomes lengthy, it can create a lot of distortion in forecasting.

If the workplace says that the product cannot be manufactured unless information is received three months in advance, the marketing department must receive an order from the customer at least three months in advance. It must, of course, immediately process the order.

In a competitive industry, materials must be purchased even before an order for its products is received. Now, assuming that the order the company is seeking has gone to a competitor, the materials purchased in anticipation of the order will remain in storage to gather dust.

This is an extreme case. But in plain English, to have a lead time of three months means that three months' worth of products are lying idle in the company. Assuming that the company has to initiate a model change quickly, everything bought for the old model has to be discarded. They have been "sleeping" in storage for more than two months, have become useless and must meet an inglorious death in the end.

This is not fair to the workers in the workplace who have been toiling day and night to rationalize the entire process.

There is not a single factor which commends a lengthy lead time. Shortening of the lead time creates the following advantages: decrease in the work not related to processing, decrease in the inventory and ease in the identification of problems. Altogether the workplace becomes more manageable.

A few years back, we tried to measure the lead time with the engine built at our Kamigo plant. Parts were cast in the morning and then assembled to become an engine. By evening, that engine was inside a car which was driven around the Toyota Sales Head-quarters. This has been and is the lead time in Toyota.

Zero Inventory As Our Challenge

For an industry, the most desirable condition is not to have any inventory. Of course, it is practically impossible to have a zero inventory, but that must be set as a goal. What a company can do is to accept this as a challenge and try to reduce the inventory as much as possible.

Many managers will say that they have successfully eliminated one half of the inventory previously maintained, and they can do no more. This is not good enough. If it has taken so long to reduce the existing inventory by half, obviously not enough effort has been expended.

If these managers accept the challenge of reducing their inventory to zero, they will inevitably continue the following process:

• If you get to the one-half mark,
• Reduce the remainder by half, and
• Again reduce the remainder by half, and
• Again reduce the last remainder by half.

If this is done, the inventory can be significantly reduced. In the end, there may be only one or two pieces remaining.

Ask yourself this question: "Can we do our work without any goods in process in stock?" If your answer is "no, for this process, we do need one piece," then you retain only that one piece. In this way, the workplace will be trained to retain only the very essential inventory for itself.

If the process does not call for a single item of inventory, then zero inventory must be the norm. If the process calls for one piece of inventory, then that must be kept. But it must be clearly understood that the one piece is the absolute necessity. The workplace learns to know the nature of its own work through the process of inventory reduction.

Can the Workplace Respond to Change?

We often hear that after an improvement, many lifts and pallets are no longer needed and a lot of new space is created. However, these cannot be accepted as the accomplishments resulting from the improvement. The excess recovered or found still does not contribute a penny's worth of profit for the company. A proper procedure is to give this information to the planning division. After all, those items that have just been removed at one time looked as if they were all needed. The cause of this waste was the improper method of production. With this feedback, the planning division can plan better next time around.

Normally, planning is done with the existing condition as the base. Thus, if the present method is quite wasteful, all types of waste can still be included in the next plan. Once an investment is made, no amount of improvement can later recover it. This is a very serious matter.

Beware of the relationship between planning and the present status. Be on the lookout for eliminating waste in the workplace. Do not neglect to inform the planning division at all times of your discovery of waste.

The above is a description of the Toyota production system and the basic thoughts behind it. We have paid special attention to the issue of economic judgment, the manner in which we use these criteria to promote our cost reduction activities. There is one final point which requires further elaboration — that is the issue of "economy," which differs from time to time depending on external conditions.

To make a rather harsh statement, what was profitable until yesterday may be a losing proposition for the company today. For example, if the wage contract is changed from an hourly-rate contract to a subcontracting agreement, the issue of profit and loss will take on an entirely different meaning.

An important thing to remember is to remain flexible. When trying to reduce waste, conditions always vary. Your way of thinking and implementation plan for improvement must always be based on this consideration.

3

Leveling — Smoothing Out the Production System

Peaks and Valleys of Work

In a normal workplace, the more the flow of things varies, the greater the incidence of creating waste. The capacity of the workplace is often adjusted to the peak work demand and not to its average value. At Toyota, there was a time in which this was the normal occurrence for us also.

Assuming that the amount of work in a day (or a week or a month) varies as shown in the illustration, the capacity of that workplace must be adjusted to the peak demand, and it must have the requisite number of personnel, machines and materials.

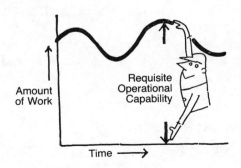

Figure 8. Peaks and Valleys of Work

However, when capacities are adjusted thus to the peak demand, under-utilization occurs when a smaller amount of work is available. If that is not the case, the worst waste arising from overproduciton may result.

The same story can be heard in the accounting division at the time books (or accounts) are closed.

In the accounting division, the peak appears within a one-month unit, or a six-month unit. At the normal workplace, the peak does not appear with such a large cycle. It may appear once every hour or once every ten minutes. We must be prepared to deal with these peaks that come in a piecemeal fashion.

This is the story of a line at a Toyota cooperative company which does the coating work.

The headlights for cars are currently made of synthetic resins. The frame of the headlight, also made of synthetic resins, is coated at this factory. Forty-eight frames are placed on one hanger for coating; previously this task was performed by five workers.

The hanger was timed to move every two or three minutes. In the case of headlight frames, one worker could hang 48 pieces in one minute. However, there were many small pieces which also had to be coated. For those that were 2 to 3 centimeters in length, 3,000 pieces had to be hung within the allotted three minutes. In this case, even with five people working side by side, it was impossible to keep up within the allotted time. The number of workers remained constant at five. Yet depending on the size of the pieces to be hung, the amount of work varied widely.

Although there was nothing unusual about the situation, it was a serious problem. We thought about this and finally decided that the space in the headlight frame could be utilized. Small pieces were hung within this space.

What we accomplished was to equalize the amount of work by "leveling" the peak. What five people could not handle before is now done by two or three.

Gift Shops in Tourist Spots

The peaks and valleys of work can be found in almost any workplace, and many companies insist on having personnel and machines to be able to cope with the peak demand.

Why do they insist on maintaining the capacity to meet the peak demand? Because they are not aware of the waste inherent in it.

Tourist spots illustrate this point clearly. Their appeal is seasonal. When guests arrive in season, there is no more space for parking, the toilet may overflow and the food served is terribly expensive but not tasty. Coca-Cola, which we buy at 40¢ a can, becomes $1.00 a can. This is something we experience very often.

From the point of view of the guests, this does not make much sense. But from the point of view of gift shops, it is natural that they want to recover the cost incurred during the entire year. They want to do so during the peak season and also add a little profit.

When not in season, many gift shops simply close down. Their proprietors may be engaged in other businesses. In the meantime, the shops, dishes and dinnerware, parking fields, etc. remain idle without raising a penny for the proprietors. At the same time, owners still have to pay taxes, mortgage payments and interest on the operating capital previously borrowed.

Gift shops and restaurants in tourist spots have no other choice but to do as they do. But manufacturers who sell their products cannot do the same. They cannot say, "We have to charge you this much because our cost has been rising," forgetting that they have wasted a lot of resources. The fierce competition existing in the manufacturing sector prevents them from taking this easy way out. This issue has been discussed in the preceding chapter.

Even among gift shops in tourist spots, if guests arrive in equal numbers regardless of the season, sales will stabilize. In such an instance, operational efficiency will be good, even without much effort.

The most efficient condition exists when the amount of work is equalized, or when the work itself is performed at an even pace.

On the Automobile Assembly Line

How does the concept of equalizing apply in the case of car manufacturing? We shall now proceed with examples from the assembly line.

If we assemble 200,000 Corona units a month and we operate 20 days, our daily allotment comes to 1,000 units.

It is easy to say 200,000 units. But even within the same Corona, specifications are extremely varied. When styles, tires, options and paint colors are all taken into account, it is possible to design and manufacture 800,000 combinations through these specifications. We manufacture them according to the orders we receive.

In the case of Crown, there are 250,000 possible combinations; in the case of Corolla, there are 16 million.

Of course, in reality we do not have that many varieties of specifications moving through our assembly lines. For Corona, that number is usually at the level of three to four thousand, and the question becomes how to manufacture these three to four thousand varieties. In other words, in this particular example, how do we line up the four thousand varieties when we assemble 200,000 Corona units a month.

A thought that immediately comes to mind is to assemble those that are similar in specifications close to one another. If the exterior paint is white, then we assemble those with white paint specifications together.

For the painting process, the procedure suggested is very convenient. All we have to do is to paint everything the same color. We do not have to clean the pipe — the procedure that is required when the paint color is changed. In fact, we do not even have to exchange the paint gun.

For the assembly process, we have to keep in mind that there are five different types of engines that can go into a car that is painted white. If by chance the same type of engine is ordered for successive cars, the work process becomes identical. There will be no mistake committed in installation, and efficiency will certainly improve.

However, in reality this cannot happen. We have experienced that in a given month, with its 200,000-unit production, if we can get 50 units a month with similar specifications, that is as much as we can expect. A more realistic view is that each car has its own specifications, and we must manufacture them accordingly.

Processes Are Linked

There are roughly 3,000 different types of parts required to build a car. If we count each bolt and each screw as a separate unit, then we will need 30,000 pieces.

Is there a better way of assembling a car using these 30,000 parts?

In the previous example, we talked about using white paint exclusively. This means that the paint manufacturer must make only white paint. Now, assuming that we want to establish manufacturing processes differentiated by color, we will need a manufacturing process for blue and another one for yellow. But if we only utilize the line with white paint, the blue and yellow lines must remain idle. As for the paint manufacturer, there cannot be an even work flow.

When the exterior paint is white, the interior often calls for either black or blue. This means that the seat lines for brown and red must remain idle. Thus the work of those engaged in car seats cannot achieve equalization either.

Behind each of these 30,000 parts, there are manufacturers and processes. We must find a way to equalize these 30,000 parts and move them forward. An assembly method that can respond to this requirement must be created.

Leveling Quantities and Types

We have discussed the waste in capacity when geared to the peak demand. Even so, when producing only a single item, it is not impossible to rearrange the production plan and personnel to level somehow the peaks and valleys of the workload. For example, a process that has less work can come to the aid of those having excess work. In this way, a single-item manufacturer can reduce waste.

However, to even out production for the automobile industry is an entirely different matter. The industry has multiple types of parts in multiple numbers. The process it must go through is a very complex one.

The only viable solution for most car manufacturers (including Toyota in its earlier days) has been to maintain a certain amount of inventory on hand. They have planned in such a way that every line will have some work to do every day. However, this approach is a costly one, because it requires holding a parts inventory three to four times larger than that required when the assembly line has an equalizing system of production. The waste created is enormous.

What is the solution then?

To have a successful system of equalized production, we must equalize not only the quantities but also the types.

In the case of the Corona already discussed, we have a production schedule of 1,000 units a day. All units are different in their engines, transmissions, axles, bodies, external colors and interiors. We scatter them all, and then do our assembly work.

Many visitors to the Toyota assembly line will ask: "Why do you have a red Corona here and another red Corona there? Why don't you bunch all the red ones together and let them flow in sequence?" The reason is very simple. We want to equalize the types.

If we allow cars with red-colored exteriors to be placed on the assembly line to the exclusion of others, red seats and interior parts will flow very heavily in the morning. In contrast, in the afternoon, there may not be enough work left for those dealing with the red color.

As for the engine, we try to let the 2000-cc and 1800-cc engines flow roughly in proportion to the number used. As for the left steering wheel cars for export and right steering wheel cars for domestic use, the determinant factor in the assembly line is the sales records of that particular time. Or we may make every third car with a left steering wheel.

There must not be peaks and valleys in our work, even in the most minute parts of the process. In so doing, we can then proceed to the equalizing system of production for our entire process.

This equalization of the quantities and the types is called *load smoothing (heijunka)* under the Toyota system. The load-smoothing system of production is the major premise for the elimination of waste.

The kanban system can succeed in a place where the final process is under the load-smoothing system of production. If there is no load-smoothing system of production, the kanban system will fail.

Cycle Time

When the factory attempts to equalize not just the quantity but also the type, what can be adopted as the standard to even out variations in the type?

In every work, timing is crucial. If not done adequately, the delivery time may be missed and the order may be canceled. On the other hand, if the product is manufactured too early, there may be a mountain of inventory. In baseball, if a runner reaches the plate just in time, he is safe. But if he is a little late, he is out.

This timing is determined by no one other than the customer.

Let us assume that Corona is sold in the quantity of 200,000 units each month. It means that 1,000 units must be produced each day (assuming that there are 20 work days in a month). In an eight-hour a day operation, 1,000 units must be produced in 480 minutes. Therefore:

$$\frac{480\,\text{minutes}}{1,000\,\text{units}} = .48\,\text{minutes}$$

In other words, one unit must be produced every .48 minutes. Otherwise, the company will not be able to meet the demands of the customer.

In this way, for every product or part, it is important to have a notion of the *cycle time*, which is defined as the minutes and seconds required to produce one item.

The cycle time is a key concept in manufacturing things. It is determined by the customer. In other words, it is determined by the sales record. The waste arising from overproduction can be eliminated through the use of this cycle time. True efficiency — not an apparent one — can result from its application.

An Example of Processing a Gear

At one section of a factory within the Toyota headquarters compound, one worker is responsible for 16 machines that process and finish a gear. This phenomenon would not be surprising at all, if all the machines did the same work, as is seen in automatic spinning machines. But in the case of these 16 machines, each has a separate function. One may grind, another may cut and shave, and so on.

Let us observe how one worker manages. First, he takes a gear coming from the preceding process and sets it on the first machine. He removes from the same machine a gear already processed and puts it into the chute. The gear is rolled over to the next machine.

The worker then moves from the first machine to the second, and while moving he turns on the switch located between the two machines. At that moment, the first machine starts moving.

The same motion is repeated at the second machine before he moves on to the third. While he walks he turns on the switch, and the second machine starts moving.

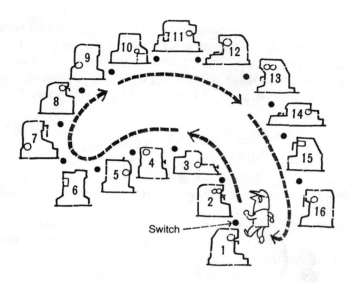

Figure 9. Processing a Gear

As he repeats the same motion over and over again, he can make a round of 16 machines in exactly five minutes. In other words, one gear is completed if a worker makes a round of 16 machines in five minutes.

Now, if we need to mass produce the gear, we can place one worker each at these 16 machines. By simple arithmetic, one gear can be produced in a little over 18 seconds.

However, if the car that uses this type of gear is only sold every five minutes — or, in other words, if the gear's cycle time is five minutes — then there is no need to station 16 workers.

In this instance, it is sufficient to have one gear every five minutes. We do not need to produce more.

Method of Load-Smoothing Production

Implementation of the load-smoothing system of production becomes easier when one has a clear notion of what the cycle time is.

We have previously stated that there are 800,000 specifications for Corona. However, in order to make the explanation of the load-smoothing plan simpler, let us assume that Corona has only five types, named A, B, C, D and E.

The required amount (amount of production) and cycle time for the five are given in the table below.

	Month	Day (480 minutes, 20 work days)	Cycle Time
A Car	4,800 units	240 units	2 min.
B Car	2,400 units	120 units	4 min.
C Car	1,200 units	60 units	8 min.
D Car	600 units	30 units	16 min.
E Car	600 units	30 units	16 min.
	9,600 units	480 units	1 min.

Figure 10. Required Quantities and Cycle Times for Five Auto Models

To obtain the cycle time, the following simplified formula may be used:

$$\text{Cycle time (tact)} = \frac{\text{Daily operating time}}{\text{Required quantity per day (unit)}}$$

Often we see lines making mistakes in obtaining the cycle time. Care must be taken to ensure that this does not happen.

The mistakes occur because these lines calculate everything from the present condition, including equipment capabilities and man-hours. They say: "We have this much equipment capability and this many people. Therefore we can produce this many units. And we can produce one unit in this many minutes."

From the perspective of the Toyota system, this approach is totally wrong. "We need this many units today," however, will give the right start. One must obtain the number of people needed to work from the cycle time, which in turn is obtained from the required amount of production for the day. Toyota's aim is to do the work required with the minimum number of people. If one calculates based on what he can do with the people the line already has, then the result is likely to be having too much capacity, creating the waste arising from overproduction.

How to Let Things Flow

Now that the cycle time has been determined, how does it work out in the actual assembly line?

Let us assume that A through E are all assembled by lines exclusively devoted to each. Thus as seen in the figure below, at Line A, units are moved at a two-minute interval, but at Line E, only one unit is assembled every 16 minutes.

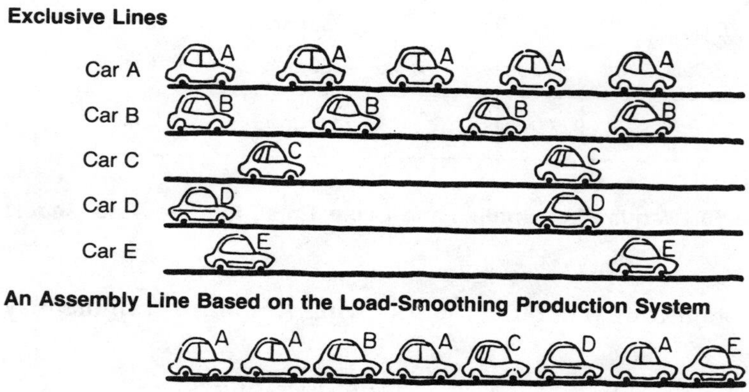

Exclusive Lines

An Assembly Line Based on the Load-Smoothing Production System

Figure 11. Load Smoothing Auto Production

When the separate exclusive lines are merged into one, the flow will take the form indicated at the bottom of the figure. At Toyota's assembly line, the cars may be the same Corona, but there are many different colors, two-door and four-door models, left and right steering wheels, all mixed together and interspersed while moving through the assembly line.

With this assembly line in operation, it becomes possible to have load smoothing not just for the quantities but also for the types. The work done in this fashion at the final assembly line guarantees that equalization can occur in all preceding processes.

Now take another look at the upper portion of the figure. These exclusive lines could just as well be processes devoted to parts (or to assembly). When all of these lines are equalized, then every line can have adequate work, and work can also become equalized.

Plan Must Also Be Load Smoothing

The load-smoothing system of production has been created to eliminate peaks and valleys in the work load and to avoid excessive production and excessive progression in a particular process. Its aim has been to equalize the work load. However, there is still another contribution that has not been discussed.

That is, through this system it has become increasingly easier to change the production plan. And the workplace has accepted plan changes more cheerfully.

One line produces 100 units every day. A plan change calls for increasing its output to 105 units daily. The workers do the job without thinking of changing their production capacity or system.

However, if suddenly an order requiring 150 units daily is sent to a line that has been producing 100 units, difficulties will arise. Overtime may become necessary, there are not enough workers, and in an extreme case, a new machine may have to be brought in. If this condition persists, the company may be forced to hire more personnel or to utilize subcontractors. Its ability to respond to change becomes rather limited.

The key to solving this problem is the production plan. But almost every factory and company seem to have poorly managed production plans.

During the month of January, the line in question is producing at the rate of 100 units. In February, it is likely that 120 units must be produced because orders are coming in. This fact is usually known by January 10. However, a common practice is to draft a plan for presentation at monthly production meetings. The workplace receives a written plan after January 20. In an extreme case, a 20-percent increase in February and an additonal 20-percent increase in March are the facts already known, but management insists on waiting until the very last minute to hold its monthly planning meetings. Management does not think to change the previously set dates for these conferences.

If management continues to follow this practice, there is no way the workplace can respond to change. It becomes a victim of its own silly rules, resulting in a curious inability to move ahead.

Therefore, when making a change in the production plan, build in the necessary changes gradually. Change the plan when an increase or decrease in the amount of production is known. In this particular example, it was known by January 10 that the demand would be higher by 20 percent in February. The management should have asked the workplace on January 11 to increase its daily output immediately by five to eight units. By taking this gradual approach, the workplace could have handled it well.

To have a load-smoothing system of production, load smoothing must also become part of the production plan.

Easier to Establish Standard Operations

In any work, it is important to establish standards. But unless the work itself is somewhat stabilized, standardization becomes rather difficult. In some instances, standards may be established, but they are worthless for practical purposes.

"The first step toward improvement is standardization." Where there is no standard, there can be no improvement.

When we engage in the load-smoothing system of production, we can establish standard operations throughout the entire process, covering all processes and lines. This is one of the main goals of load-smoothing production.

At Toyota, we manufacture through load smoothing, we figure out the cycle time and we create standard operations. We then promote our improvement activities. These are the basic steps we have consistently followed.

In a nutshell, the Toyota kanban system removes the signboards proportionate to the quantity used, and goes to the preceding process to withdraw exactly the same quantity. The preceding process manufactures the exact quantity just withdrawn. The kanban system is a system that keeps on turning this cycle.

Frequently we hear this comment: "That's a very snappy way of doing things. Even for those parts bought from outside suppliers, all one has to do is send a piece of paper and the parts arrive." But the success of the kanban system is dependent on the thoroughgoing conformance to the load-smoothing system of production at each of the final processes.

Assuming that the final process has not been converted to the load-smoothing system of production, but insists on using the kanban to receive parts — for a party receiving this demand, it is a bolt from the blue.

The workplace cannot use the kanban to order 50 boxes today, none tomorrow and 150 boxes the day after tomorrow. If the preceding process or supplier is treated in this manner, utter confusion will result.

Under the kanban system, the subsequent process withdraws parts and materials from the preceding process every day, with consistency — in the same manner, at the same interval and in about the same amount. It is only in this way that the system can succeed.

Obstruction Called Exchange of Die

A bottleneck is often created in the load-smoothing system of production by the exchange of die.

Normally the exchange of die is considered a time-consuming process. Why?

The preponderant reason is that where there is no desire to do die exchange quickly, it becomes a self-fulfilling prophecy.

In fact, some workplaces do not worry if the exchange of die takes eight hours. That, of course, is an extreme case, but no one thinks much about the die exchange that takes one hour. In some workplaces, equipment procurement is often premised on these uncontrolled die-exchange procedures.

In places like these, the load-smoothing system of production and the exchange of die become irreconcilable opposites.

One of the characteristics of the Toyota system is to make the size of a lot in the workplace as small as possible. If the time spent for exchanging die is greater, the lot is likely to remain larger. When the lot is larger, it is often assumed that through it the time lost through the exchange of die can be recovered. But this can lead to the waste arising from overproduction.

Our aim is to please our customers by producing only those cars that they order. If we complain about the frequency of die exchange, we are in effect saying to our customers: "Why don't you order the same style and same type of cars?" That is not the way to do business.

The only alternative left for us is to shorten the time spent for the exchange of die.

Preparation and Cleanup Are the Keys

It is not that difficult to shorten the time needed for the exchange of die.

The point is to make an advance preparation of those molds, jigs and tools that can be assembled ahead of time, and to clean and place in their storage places those molds, jigs and tools that are removed only after the machine starts moving again. This is called *thoroughness in the outside exchange of die.**

We must concentrate on, and try to improve, those operations that cannot be performed without stopping the machine. With this alone, the time needed can be significantly reduced. This is called *thoroughness in the inside exchange of die.**

If tools are to be used, be sure that all those needed are kept by the side of the machine and in the order they are to be used. Do not overlook the proper arrangement of materials. Focusing their attention on the die exchange, workers can perform it quickly and expertly. But this can come to nothing if they do not have the necessary materials. Things like this happen frequently.

Preparations and arrangements can be made through improvement in the work process. In other words, these can be accomplished through the workers' own ingenuity. The procedure must be standardized and written down in a standards manual with which workers can continue training and retraining to do the exchange of die. Time reduction becomes possible.

This training is very similar to fire drills practiced by many companies. In a fire drill, it is not unreasonable to expect that all the water hoses are readied within two minutes. Everyone observes the correct procedure. The fire fighters also know how to divide up their work without any waste or delay.

In exchanging die for large equipment, Toyota retains the services of a special unit organized by specialists in die exchange. Seven or eight years ago, it took three hours to exchange die in an eight hundred-ton press. Today it takes only three minutes.

---

*   See *A Revolution in Manufacturing: The SMED System,* by Shigeo Shingo (Productivity Press, 1985). Mr. Shingo, who developed the quick die change method for Toyota, explains inside exchange of die (IED) and external exchange of die (OED) in detail, and outlines steps for reducing and streamlining both IED and OED.

New Ideas Help Shorten the Time

Generally speaking, the time spent in the exchange of die can be broken down in the following manner: preparation, 30 percent; removal and mounting, 5 percent; centering and setting the measurement, 15 percent; and adjustment and trial processing, 50 percent.

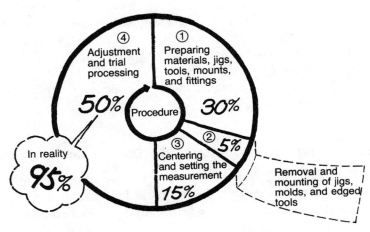

Figure 12. Time Needed for the Exchange of Die

As discussed earlier, the time spent can be reduced considerably by differentiating between the inside exchange of die and the outside exchange of die. The machine must be stopped for the inside exchange. But when this procedure is externalized as much as possible in an orderly fashion, time reduction becomes a reality. The worst enemy for time reduction is the final procedure of adjustment and trial processing .

Often the time spent for adjustment and trial processing is not 50 percent but closer to 95 percent of the time needed for the die exchange.

To exchange die means to start everything anew, even though the previous setup functioned smoothly, it was well centered, its measurements were determined, and it was processing effectively. Now a different mold is installed, and everything has to be done over again, including the centering, determining of measurement and adjustment. In a trial run, however, there are a number of defectives, and the entire process may appear to be a losing proposition.

Here we need a new way of thinking. We must discard the notion that the exchange of die is a time-consuming process, as well as any other established notions. No matter how many times we exchange die, the important thing to remember is that we restore the original correct state. This is just like playing golf, where swinging is not as important as having the correct form.

Using this notion of "restoring," when we take another look at the exchange of die, it becomes merely an exercise in placing something that is moving at a given location or position onto that which is not moving.

If this is done correctly, we do not need the cumbersome process of adjustment.

Devising Plans to Shorten the Die-Exchange Time

By changing our thinking process, we have been able to devise a number of steps at Toyota to reduce the time needed for the exchange of die. Several of these steps are described below:

Color coordination: Each time we exchange die, we use a lot of tools and bolts to tighten or dismantle. We differentiate these tools and bolts by color. When we do this, mistakes are eliminated and operations become faster and easier.

Bolts may be different in size, but if we make their heads uniform, we do not have to change our tools.

Color coordination is also utilized in the different types of hose we use, in the LS dock and in pressure-adjustment handles.

Preheating: We utilize the remaining heat of the retaining boiler attached to a die-cast machine to preheat the next die. In this way, we not only save time, but also reuse energy.

Press dies: There are many different types of press dies. By unifying their heights, adjustment of strokes becomes unnecessary. By using jigs, installing stoppers and cutting key grooves, the process of adjustment can be eliminated.

Do not use crane: For the removal of dies weighing up to 2.5 tons, we do not use cranes or forklifts. We create a flatcar which allows us to push or pull by hand. During the preparatory stage, however, cranes may be used.

Use of auxiliary jigs: To install dies into a bolster or cutting tools directly to a chuck takes time. Therefore, at the stage of outside setup, they are placed on an intermediary jig ahead of time. During the inside setup, it can be installed on the machine promptly. In centering a drill bit, we use the same procedure. Both are effective and time saving.

A Manufacturing Section Chief's Setup Guidelines

Reproduced below are guidelines prepared by the chief of a manufacturing section for shortening the setup time. They serve as a useful checklist.

Objectives of Shortening the Setup Time

- Making work easier, simplified, more easily managed
- Making work safer
- Stabilizing quality, standardization, foolproofing, lowering the cost and inventory

Procedure for Setup Work

- Is the procedure standardized?
- Are there waste *(muda)*, unevenness *(mura)* and unreasonableness *(muri)* in the work content?
- Do you know which work content is truly necessary?
- Are the dies, tools and gauges that are needed prepared at the stage of outside setup?
- Are the needed materials within reach?

Points to Consider in Shortening the Setup Time

- Is there any part removed unnecessarily?
- Are all appropriate tools available?
- Can you reduce the number of types of tools?
- Why is adjustment necessary? Can that process be eliminated?
- Can you eliminate the use of some of the bolts?
- Can you make any of the processes into a one-touch operation?

- Which is better — exchanging parts or exchanging sub-assemblies?
- Can this process use gauges or spacers?
- Can the contents of this setup be simplified through changes in reserves, processes or designs?

One-Shot Exchange of Die

There are many machines standing side by side. The process calls for bending, stamping, welding and boring. In this multi-process equipment situation, how can the die exchange proceed?

Assume that we are to process parts A and B in sequence and we have four machines. We do *not* process part A at all four of these machines first, and then when this process is over, exchange all the

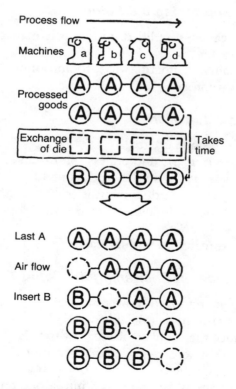

Figure 13. One-Shot Exchange of Die

dies in these four machines to start processing part B. If we did this, it would simply take too much time.

An alternative method which is used by Toyota is demonstrated at the bottom of Figure 13. Regardless of how many machines are connected to each other, a processed part flows one by one with each cycle time. Therefore, we do not allow anything to flow after the very last A. Instead we let the air flow.

While the air is flowing into each of these machines sequentially, we engage in the exchange of die. In other words, our setup change takes place within one cycle time. At Toyota, we call this a *one-shot exchange of die*.

Since we use only one cycle time for the exchange of die, we only encounter a loss of one part piece.

4

Just-In-Time and Automation

Supermarket Style

We have discussed the basis of the Toyota system, the load-smoothing system of production, in the preceding chapter. In this chapter, we shall study the two pillars standing on this foundation, namely the just-in-time system and automation with a human touch.

Just-in-time means to supply to each process what is needed when it is needed and in the quantity it is needed.

This is not unique to Toyota. Whenever there is a production plan, it means that the company is seeking to attain just-in-time. They want to do their work systematically; remove excess work; eliminate waste, unevenness and unreasonableness *(muda, mura, muri)*; and raise their productivity.

In the decade immediately after the Second World War, Toyota always had a production plan ready in the first part of the month. But parts could not be gathered until the middle of the month. As for the assembly line, most of its work was done near the end of the month. In this way, the company somehow met its monthly quota.

It was an exasperating situation. No matter how hard they tried, the assembly lines could produce only one car for every ten workers.

Around that time, Mr. Taiichi Ohno began the study of supermarkets. His thinking was that the structure of a supermarket could somehow be made applicable to a manufacturing plant.

At a supermarket, each customer selects the type and quantity of foods he needs from the display cases, puts them in the shopping cart and pays for everything at the counter. He then takes his purchase home. In making his purchase, he takes into account the number of people in the family, what space is available in the refrigerator and the number of days the supply must last.

Mr. Ohno's discovery was that perhaps this system — to take home (buy) the items needed in the amount needed — could be utilized in the place of production.

In those days, Japan's delivery system was archaic. One may wish to have *sushi* for only one person. But it does not sound right to ask for home delivery of just one order, so he orders *sushi* for two. In asking vegetable vendors and *sake* merchants for delivery, the same consideration prevailed. A small quantity does not sound right, so an excess amount is always ordered. From the point of view of a customer, the supermarket-style purchase ensures that there is not going to be an excess purchase. One simply goes out to buy what is needed when it is needed.

The term *just-in-time* was invented by Mr. Kiichiro Toyoda, the first president of Toyota. But it was Mr. Ohno who took up the challenge. Indeed, Mr. Ohno is responsible for the creation of the Toyota system as we know it today. Let us now hear from him in his own words.

Just-In-Time

"When I first started working for Toyota, I heard that parts were to be assembled 'just in time.' I thought it was an interesting expression, but in reality it was not done. I began thinking about a good way of implementing this just-in-time concept. Maybe I am a perverse sort of a person: I have a habit of reversing the process in my thinking. Using this process, I began to think that all we had to do was to *let the process that needed parts go to get what was needed, when needed and in the quantity needed*. In this, I simply reversed the system of transport.

"Previously in every company — and we were no exception — as soon as one process completed its work, its products were sent to the next process. In some instances, subsequent processes served as intermediate warehouses. There were all sorts of things, but altogether it was the responsibility of the process that manufactured parts and products to carry them to the next process. My thinking was that the responsibility for transporting should be given to the subsequent process. As for the preceding process, it simply had to put down what it manufactured. When the subsequent process needed these products, they had to come for them, but only when they needed them.

"As the subsequent process came to the preceding process to receive materials, the preceding process had to replace whatever had just been transported. By adopting this method, intermediate storage areas became unnecessary. The preceding process manufactured what was needed and stored the items. Once the storage place was full, production had to stop.

"This system has a number of advantages. Assuming that there are many workers, and machines and equipment have excess capacities, some managers may feel that it is wasteful to let them remain idle. So they keep on manufacturing. The next thing they know is that they do not have enough storage space. That is why I tell them that they must keep whatever they have made in the place where they have made them. They must replace what has been transported elsewhere. In this way, all workers can find out what must be made simply by looking. Workers will know whether they are too slow or too fast. Now, suppose there are materials in abundance, but since there is no place to store the item manufactured, workers are forced to remain idle. When that happens, both the supervisor and workers will know that perhaps the process does not need that many people. In this way, placement of people becomes a relatively easy task. By creating a transport system which has gone in the reverse order, just-in-time has become a viable, practical method."

Withdrawal Is Made by the Subsequent Process

Automobiles consist of tens of thousands of parts. The processes required to manufacture a car are extremely large in number. To harness all of these different processes to engage in the just-in-time system of production is a very difficult task. To reach this goal in reality, production plans have to be changed rather frequently.

Factors causing changes in production plans are numerous,

including changing market conditions and varied factors affecting the manufacturer. When these factors are at work, and if a problem arises in the preceding process, the next process may find a shortage in parts and in other items. Whether they like it or not, some subsequent processes may have to stop their lines or change their plans.

If management disregards these existing conditions and shows to each process an ironclad production plan, it can create several undesirable consequences. Among them are (1) producing parts without regard to the needs of the subsequent process, and (2) creating a serious shortage as well as a stockpile of unneeded parts, all at the same time. The ironclad plan still has to be changed and causes difficult problems in the process. To give new instructions and to make adjustments alone can take up many management man-hours. Even if the task of management is handled adequately, there still remains the mundane matter of cleaning up, rustproofing and counting. All of these can be translated into substantial man-hours. It is a sure way of creating a hotbed of waste in the workplace.

There is another, even more undesirable impact. At each of these lines, workers may lose the sense of what is normal and what is abnormal, and treatment of abnormal conditions may suffer as a result. There are too many people, and the line overproduces. In this case, if the situation is known, improvement can be implemented. But people can become victims of the plan and nothing is done. When these factors interact, wastes are created one after another. This in turn can become one of the reasons for the worsening of the company's economic condition.

However, if a process can supply items that are needed at the time needed and in the quantity needed by other processes, the waste as discussed above can be eliminated from the workplace and improvement can move one step forward. To accomplish this, management must abandon the practice of giving production plans to each and every one of the processes, or of making the preceding process transport its products to the subsequent process. This system does not make it clear how much the subsequent process really needs, and the preceding process is liable to overproduce. Productivity suffers when a process produces an unnecessary amount or transports to the next process materials not needed by the latter.

A new idea is born to reverse this process. It is done with the subsequent process withdrawing from the preceding process what it needs. Instead of the preceding process sending to the next process what it has produced, the flow is changed in such a way that *the subsequent process withdraws from the preceding process what it needs when it needs it. The preceding process produces exactly the same quantity as is withdrawn.* In this way, all of those variegated problems mentioned earlier can be solved.

The final point of manufacturing processes is the final assembly line. This is made into a point of departure, and a production plan is shown only to the final assembly line. In it is shown what types of cars, and in what quantity, are needed, and when they are needed. The assembly line goes to the preceding process to withdraw parts it needs. In this way, steps are taken to retrace the entire manufacturing process backward, and even the division of rough raw materials can become simultaneously connected with the rest of the manufacturing process in a chain-like fashion. Conditions to fulfill the requirements of the just-in-time system are satisfied, and the management man-hours can be significantly reduced.

The kanban is used at this time to withdraw parts or to order the production of parts. Through the kanban system, just-in-time production can be performed smoothly, and waste in the workplace is eliminated substantially. It reaches a nearly ideal situation in production management. In addition, lines become more flexible, putting a brake on waste-making.

Automation with a Human Touch

Another pillar of the Toyota system is *automation with a human touch*.

There are many machines that start moving once the switch is pushed. We also have many high-performance and high-speed machines. If something unusual happens, e.g., if a foreign substance enters, the equipment and die can be broken. Scrap may clog up the machine's regular process, or if the tap is broken and defects begin to appear, it can immediately create a mountain of tens or hundreds of such defective products. In such a case, machines have not worked, nor have they functioned properly. How can we prevent it? Do we assign a watchman to each of the machines? If automation means precisely that, then one cannot expect efficiency through automation. At Toyota, we strictly forbid this type of automation.

Let us now hear from Mr. Ohno again, to see what he says about automation.

"At Toyota, we insist that automation be accompanied by a human touch *(ninben no tsuita jidoka)*. Without this human touch, automation loses its meaning. Any machine can be automated, and other manufacturers can have automation with such machines. But we add the human touch, and as users of these machines, we add that touch to them.

"In brief, this automation with a human touch has an automatic stopping device if something goes wrong. When the processing is over, or a defective item is manufactured, without this automatic stopping device the situation can get serious. If defects were created in a large quantity, it would become difficult to control. We simply had to install a device which could prevent mass production of defective items."

Add the Wisdom of Those Who Use It

"This *automation with a human touch* is a phrase that was probably invented by Mr. Toyoda also. The late venerable Sakichi Toyoda, the founder of the Toyota group, created an automatic loom with a human touch.

"In weaving, strict standards are established. Within one square inch, an exact number of threads for warp and woof is required, and with that specification, the name of the weaving is firmly set. If one thread of the warp or woof is missing, it becomes a defective product.

"Toyoda's automatic loom stopped immediately whenever a thread of the warp or woof was snapped, or whenever the thread was no longer in the loom. In other words, the loom could not produce defectives, because it had an automatic stopping device. We were told that this was a machine with a human touch.

"When I visited the Toyoda Automatic Loom Manufacturing Factory, I teased them: 'Say, your factory used to be called an automatic loom manufacturer with a human touch. I see that you have removed that human touch from your name. Are you getting poor?' To have a human touch means that the automated machines can monitor themselves when defectives are produced. These machines stop just before a defective product is about to be manufactured. After this invention, it has been possible for one worker to handle 20 or so looms, and these looms move at a very high speed. Compared to the days when an individual loom was controlled by one's feet, productivity must have risen exponentially.

"We have applied this thinking in making automobile parts. By adding that requisite human touch, we are thinking that it is not impossible to raise our productivity ten or even 100 times.

"We say to our employees:'You have just bought these automatic machines, and they still do not have the human touch. That human touch must be provided by you. By human touch we mean that we want you to impute your knowledge to the machines you handle. If you simply operate the machines just bought, you do not show any particular ingenuity. To make your daily work worthy of you, put your touch to the automated machine.'"

To Stop Automatically

While no one today will challenge Toyota's claim as a world-class car maker, the situation was quite different in 1937 when Toyota was founded. It had to meet head-on with competition from the more advanced American and European auto makers. To catch up and to overtake the competition, it was necessary to automate its equipment and facilities. From 1955 until 1965, the company took a giant stride toward automation.

However, with automation, manpower requirements did not seem to decrease. In some extreme instances, there were rows of automated machines, but an inspector had to be placed by each machine. In effect, these machines were no different from manually-operated machines.

This does not mean that a manually-operated machine was desirable. At that stage, we simply did not have an adequate perception of what automation ought to be. The first step toward automation was not how the machine should process automatically, but rather how it could detect abnormalities and stop automatically.

What we needed was not an "automatic" machine, but one with a self-sensing human touch. At Toyota, we put forth a lot of effort to make machines, old and new alike, have this self-sensing human touch through our human ingenuities. Such examples can be found in the system of making the machine stop exactly at a predetermined spot, in the full-work system, in the foolproofing system, and in various other safety devices.

This "automation with a human touch" concept is extended beyond the machines to the places where people work on the assembly line. People and machines must stop promptly if an abnormality is found. This entire system is what we call "automation with a human touch."

The Japanese term for automation is *ji-do-ka*, consisting of three Chinese characters. The first character, *ji*, means the worker himself. If the worker feels that "this is not right," or "I am creating a defective product," he must immediately stop the conveyor. We are suggesting that each worker, in effect, has a stop switch for the line. Whenever he feels that something is wrong, he immediately stops the line.

SAYINGS OF OHNO

Automation with a human touch means that a machine can stop itself by using its own judgment. Automation without a human touch is merely an ability to move.

Automatic machines (without that crucial human touch) may destroy machines and dies where there is an abnormality. They can produce a large quantity of defectives, and require the posting of inspectors.

Stop the Line

It is not easy to stop the line when work is in progress. When the line is stopped even momentarily, the amount of production will fall, and no supervisor cherishes the thought. At Toyota, lines are stopped — but this does not mean we like doing it.

In any event, Toyota's lines stop now and then. When they stop, they stop only for a few seconds. This is done in order to make the entire line function smoothly. We stop the lines with the ultimate goal of never stopping these lines again.

This is a true story:

There was a supervisor at Toyota, whom we shall call Mr. A, who steadfastly refused to stop his line. Then there was another supervisor, whom we shall call Mr. B, who would do as suggested in stopping the line.

Mr. B did not hesitate stopping his line, so initially it stopped too frequently. The number of cars assembled on this line was drastically cut and no production plan could be maintained. The line stopped, however, because of the many problems which, while known to the workers, had never been brought to Mr. B's attention.

By stopping the line, the problems were made obvious. They were solved one after another. After three weeks, the situation reversed. Mr. B's line outperformed that of Mr. A. The latter kept on thinking that to stop the line meant lowering efficiency and bringing a loss to the company.

It may sound paradoxical, but for us to stop the line means to ensure that it will become a stronger line which will not have to be stopped for the same reason again. The goal is to create an ideal line, and the company is willing to bear the loss in stopping the line. Thus, when the line is stopped, supervisors scramble to solve the problems.

Supervisors who can never say "Stop the line" are failures. Equally at fault are those supervisors who stop the same line two or three times for the same reason.

SAYINGS OF OHNO

The line that is not stopped is either a tremendously good line, or an absolutely terrible line.

Most lines cannot be stopped because there are so many people surrounding them, obscuring the problems.

What is needed is to make sure that the line can be stopped. And then continue to improve the line in such a way that it becomes a line that need not be stopped, even if one wishes to do so.

We must think carefully about the meaning of this phrase: "The line that is not stopped is either a tremendously good line, or an absolutely terrible line."

Easy-to-Observe Workplace

The way to stop the line is to give each worker on the line a stop button, and let him continue to engage in his standard operations. If within his work area the work is not likely to be completed, he pushes the button to stop the line.

The work may be delayed because the parts are not correctly assembled or are defective and cannot be fitted. These reasons are studied carefully, and mistakes are corrected thoroughly as they occur. In this way, the same mistake is not repeated. In the long run, it is more advantageous to stop the line.

At Toyota Motors, there is a stop button for every line. When a new worker enters the line, the first lesson he receives is how to stop the line.

If for any reason the line is stopped, a display immediately appears on a board suspended above each line, showing at which process of the line a problem has occurred. This display board is called the *andon* (display lamp).

Figure 14. The Easy-to-Observe Workplace

For example, a line consists of twelve processes. For some reason the line is stopped at process (4). The display light will immediately light up on position (4) to indicate that the problem is at that location. A supervisor nearby comes immediately to study the problem and provide a technical solution.

By concurrently using the stop button and the display lamp, conditions existing in the line become obvious with just one glance. It makes it possible for us to engage in *visual control*.

Visual Control

We have repeatedly stated that the most important aim of the Toyota system is the thorough elimination of waste. Yet it is difficult to recognize what waste really consists of. In contrast, it is not too difficult to determine the methods or ways of eliminating waste.

Thus, if we can make it obvious to everybody what waste is, we have taken the first positive step toward its elimination.

The president of a certain cooperative company came to Toyota one day and said: "We don't have enough work. Please do something." Toyota was alarmed and sent Vice President Ohno and his staff to visit its factory.

Ohno came to the factory scared. "If there is no work, the factory may be awfully quiet. I hate to see that." But what he and his staff saw was a factory with workers moving about busily and machines in full operation. There was no indication that the amount of work was insufficient.

The group was puzzled. They went there feeling that it was Toyota's responsibility to provide work for its subsidiaries, but they did not expect to see what they saw.

Ohno and his group half expected that on their visit they would be greeted by the pleading eyes of the employees, from the president down, saying: "What are you going to do about this?" But that was not to be the case.

When the group finished their study, they found that the amount of work was not as much as the subsidiary was capable of handling. But if that were the case, it would have been better to have some visual evidence. Some workers could have remained idle there.

"Have Time on Hand"

The example that follows may be a bit extreme. But at Toyota we say to our workers, "Have time on hand." When there is no work, we do not want them to do something unnecessarily. It is better for them to remain where they are and do nothing.

Workers are forced to have time on hand because they are only assigned that much work. The fault is in the assignment and cannot be assessed against the workers. When they have time on hand, their group leaders and foremen will know the situation immediately. This provides an opportunity for the group leaders and foremen to reassess the manner in which they give assignments and to initiate improvement activities.

This is another example of an easy-to-observe workplace where visual control can be effective.

Often the president, division director and factory superintendent wander around the factory floor. The workers may feel that making the rounds once a year is not good enough. Such a visit can be made into a more useful one if the workplace organizes itself in such a way that any visitor can understand at a glance what is going on there.

SAYINGS OF OHNO

Make your workplace into a showcase that can be understood by everyone at a glance.

In terms of quality, it means to make the defects immediately apparent. In terms of quantity, it means that progress or delay, measured against the plan, is made immediately apparent.

When this is done, problems can be discovered immediately, and everyone can initiate improvement plans.

The Visual Control Method

At each of the workplaces, the following steps are implemented to promote visual control:

1. ***Determine the locations*** where products and parts are to be stored and display the locations clearly. Mark the location on the kanban. Through this, abnormalities can be found immediately regarding storage control, procedure for handling work in progress, progress status and transport operations.

2. ***Erect a display lamp (andon)*** to be used in stopping the line. It will show the movability of the line, places where ill-fitting by equipment occurs, etc. Measures taken to correct them can become readily known.

3. ***Place a kanban above the line.*** This shows what work is in progress, the status of the preparation for the next stage, the relative condition of this line's load and the necessity for ordering overtime.

4. ***Display the kanban.*** Through this, the cycle time, work procedure, standard stock on hand and so on can be known.

Visual control at the workplace makes it possible to let the machines operate automatically when conditions are normal, and lets workers engage in abnormality control when abnormalities occur.

Visual control is an important concept. It is directly connected to the two pillars of just-in-time and automation with a human touch.

Cowboys Engage in Abnormality Control

Control is a word used everywhere. But what is the essence of control?

A cowboy moves a herd of cattle several hundred miles away. That is the essence of control. As Westerns show, tens of thousands of cattle are moved from one location to another by a very small number of cowboys.

Under normal circumstances, cowboys do nothing and merely follow the herd. But if the cattle move off course, cowboys move to the head of the pack and correct the course. If a few animals do not stay with the pack, cowboys use their lassos to restrain them and return them to the pack.

Assume that a regulation is issued saying that for each steer there must be a cowboy who can ensure that it will move in a straight course. In such a case, no herd could ever traverse tens of hundreds of miles across the desert to reach the destination. More likely than not, eventually all the cattle would have to be eaten by the cowboys, and the herd would disappear. At the destination there would still be a group of cowboys, but without cattle.

In other words, control is not needed if things are moving along smoothly. Only when there is abnormality, is there a necessity to move quickly to take action.

Looking ahead to what we must do, we have concentrated our control effort on those abnormalities. And at Toyota we call it *abnormality control*.

With this abnormality control, the control capability or the circle of control can be enlarged. One worker can man a number of automated machines. One group leader or foreman can observe and supervise a number of lines. Parts foremen in the engineering department can handle an extraordinarily large number of parts and remain efficient.

5

Workplace Control Through
the Kanban System

Toyota's Production Plan

"Does Toyota have a production plan?" This is a question often asked. Those who ask it reason that with the just-in-time system to make cars, Toyota does not need a production plan. To make that which is needed, at the time needed and in the quantity needed somehow appears to be a haphazard way of doing things.

However, Toyota Motors, like any other company, does have a production plan. A long-term plan, annual plan and monthly plan are formulated, derived from the overall corporate policy. A daily plan is also established, showing how many units of what type are to be manufactured daily. Naturally, the daily plan gives support to the notion of load smoothing. For example, if the monthly plan calls for production of 10,000 units and the number of days in operation comes to 20, the daily production will be 500 units. Of these 500, Style A will be 250 units, Style B will be 200 units and Style C will be 50 units.

The daily plan is then developed into a daily procedure. It shows the order in which the flow of different styles will occur. First A is introduced to the assembly line, next comes B, A comes next, and then comes C, and so forth.

However, one of the most distinctive features of the Toyota plan is that the plan for the daily procedure is given only to the final assembly line and no one else in the workplace receives it.

To the preceding processes, such as the processing process and the rough materials process, only a rough estimate of what the process needs for the month is given to each. The quantity needed in a month is, of course, not a fixed figure. But it can provide a yardstick for the operations of each process. Once the anticipated monthly quantity is known, each day's output can be calculated, and a plan for making each piece at what speed (e.g., how many minutes and how many seconds) can be established.

Thus, with the exception of the final assembly line, the workplace does not receive anything resembling a production plan chart. In a sense, we do not have uniform production instructions. This is why some may mistakenly think that Toyota does not have a production plan.

Plans Are Made to Be Modified

Why do we do things this way?

This is so because we recognize that plans are made to be modified. No matter how precisely one establishes a plan, the market condition changes hour by hour. If sales fall, the company must reduce the number of units produced. If sales rise, of course production must be increased.

If a production plan is a fixed one, it becomes difficult to change quickly. To give a measure of flexibility, some companies may continue to produce the exact quantity scheduled for this month, and make the changes required the following month. Of course, in this case, too, the production plan is modified in accordance with the sales figures.

Under a fixed production plan, information is given as a unit to a group in the workplace. Thus, even if there are troubles in the preceding process and subsequent process, the process in between keeps on producing what it is assigned to do. As a result, it creates a surplus on the one hand, and a shortage on the other. What it has accomplished is totally different from what the original production plan intended. Confusion becomes a common denominator in this instance.

The production control system that feeds only a fixed amount of information experiences this type of difficulty.

Before we go any further, there is another consideration we must not overlook. When things do not flow as planned, there are some internal factors that prevent the plan's proper execution. These factors create defective products and cause malfunctioning of the machines and shortage of parts.

A production plan is influenced by internal and external factors. Marketing conditions also change constantly. To respond to these changes, instructions given to the workplace must be modified constantly. Because we want to produce things just-in-time, production orders must be given frequently and in a timely fashion. To the workplace, the most important information is how much of which product it must produce now. It does not require a neatly put-together instructional memo. Orders can be given as needed.

Suppose the company produces 5,000 units of one type of car, anticipating a corresponding demand. But the actual sales have reached only 3,000 units. With all that work, the company is suddenly saddled with 2,000 unsold cars in storage.

In this case, a mistake was committed when the figure of 5,000 cars was adopted as one unit. Had the company considered 500 cars, 50 cars or even 5 cars as a production unit and only produced that much each time in a small lot sequentially, it could have stopped its production when it became known that only 3,000 units would sell.

Production, therefore, must be in a small lot. Continue that process. This is a sure way of preventing overproduction and leftovers. Toyota's fundamental thought lies in this concept.

Providing Information by the Minute

What is the best way to give production orders frequently and diligently?

The most important thing to know at the workplace is what must be done next. In giving these orders, if this crucial information is transmitted, it is often sufficient.

At the assembly line, workers must know what type of car comes first, then next, and so on. If it takes a minute to assemble a car, the order must be given in one-minute intervals. If a car is assembled every three minutes, then the next order should come in three minutes.

To provide timely information simply means that the manner of transmitting instructions must be consistent with the cycle time.

From the point of view of office work, it is much easier to give orders by the hour, or by the day, instead of by one- and three-minute intervals. This is the reason why information is given in a batch. But we must keep in mind that no matter how complex it may be for office workers, they should not be allowed to take an easy way out and contribute to the waste arising from overproduction.

Is office work that complex? The actual act of building a car and the act of ordering a car to be built cannot be compared. It is much easier to say something than to do it. Giving instructions is much easier than building a car.

To provide information by the minute also means that the company is engaged in abnormality control, showing how to act when something unusual happens. Toyota is not a company that would let the final product emerge without an overview or plan for it.

If there is no variance, the procedural plan that has been formulated by the computer will be followed meticulously. Based on this plan, and consistent with the cycle time, instructions are given which show what comes next in the line's production.

As a means of providing information keyed to the cycle time, the final assembly line uses the inter-writer, while most other processes rely on the kanban.

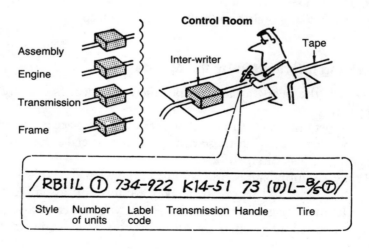

Figure 15. The Inter-Writer

The inter-writer is a type of electric communications device. At the control room, an operator follows a daily assembly procedure chart designed by a computer. He transcribes manually on a tape each car's style, tire, transmission, etc., and sends the information to the assembly, engine transmission and car frame processes. Through this electrically transmitted tape, each of the processes knows instantly what is to be made next.

The label code shown in Figure 16 refers to the types of labels that provide detailed specifications for cars to be manufactured. At the assembly line, on the basis of the label code shown on the inter-writer tape, a label prepared in advance is posted on the car body, and assembly follows the specifications given on this label.

AOI				
Assembly no.		File no.		Copy status
(Inter no. th unit.)		Destination	(Export car must use English plate)	
Car style	*BJ 43L – KJW*			
Rear spring	Rear axle	Booster	Steering lock	Collapsible handle
	Semi	*Single*	*Yes*	/
Define gear ratio	Free wheel fabrication	Electric system	Exhaust	Transfer
411	/			*Direct*
Alternator	Air cleaner	Oil cooler	Heater & air conditioner	Front winch
480 w	/	/	*Heater*	/
Cold-climate oil	Altitude compensation	LLC	Fan	Rear hood
	/		*tempered*	/
EDIC				Cold-climate destination
yes				

Figure 16. An Example of a Label

In giving directions based on this daily assembly procedure chart, unforeseen troubles can happen in some of the processes. Or changes may be required in the assembly procedure due to changing circumstances. Since the tape is handwritten, it can be altered to give new instructions whenever a change occurs.

Inter-writers are used in this fashion by assembly lines. The kanban plays a similar role for other processes, such as the forging, casting and sub-assembly processes, which constitute the overwhelming majority of processes in car manufacturing. The purpose remains the same, that is to provide timely information as a means of restraining overproduction.

The next illustrations show two types of kanban actually used by Toyota Motors. Figure 17 is an in-process kanban, and Figure 18 is a parts-ordering kanban for subcontractors. Each of these is about 3½ by 8 inches in size, and is encased in a vinyl pouch.

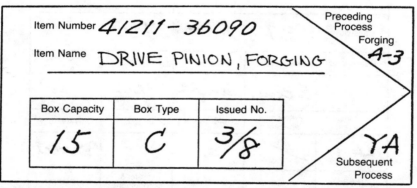

The preceding process is a forging process, and this process goes to its Section A-3 to withdraw parts. The capacity of a parts box is 15 pieces, and the box type (or shape) is C. There are eight sheets of this kanban that have been issued. This particular one is No. 3 of the eight issued. The characters YA stand for the tempering process.

Figure 17. An In-Process Kanban

The shape of the kanban is not fixed. In some processes, they are made of iron and are larger in size; others are triangular in shape.

It matters very little what shape a kanban takes. Each process and each factory can determine it. The important consideration is how best to transmit the requisite information accurately.

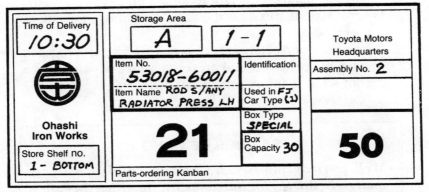

When the Ohashi Iron Works delivers parts to the headquarters factory of Toyota Motors, they use this parts-ordering kanban for subcontractors. The number 50 represents the number of Toyota's receiving gate. The rod is delivered to storage area A. The number 21 is an item back number for the parts.

Figure 18. A Parts-Ordering Kanban (Subcontractors)

Kanban's Various Functions

The kanban contains information that serves as a work order. This is its first function. In short, it is an *automatic directional device* which gives information concerning what to produce, when to produce, in what quantity, by what means and how to transport it.

The quantity to be produced, the time, methods, procedures and amount to be transported, where it is to be transported, the place to store it, the means of transportation and the type of containers to be used can all be known at a glance through the kanban.

Normally, companies give information dealing with "what, when and how much" to the workplace in the form of memoranda containing a chart for the start-up plan, a transportation plan chart, production orders and delivery orders. Information concerning the method of production, transport destination and storage areas is contained in a book on standard operations, which is often buried under a massive pile of paper on the corner of someone's desk. Seldom are these standards honored — one of the major reasons that defective items are produced.

The kanban system was created to do the following:

1. Engage in standard operations at any time
2. Give directions based on the actual conditions existing in the workplace
3. Prevent addition of any unnecessary work for those engaged in start-up operations, and prevent a deluge of paper which cannot serve as future source materials

---

SAYINGS OF OHNO

Well, you may think you are creating useful source materials (shiryo) for the company. But often they are turned into a meaningless pile of paper (shiryo) or, worse still, dead weight (shiryo). The true shiryo can have only one meaning, and it must be the true source materials.

The second function of the kanban is to move with the actual material. We have already suggested that the kanban is a tool for visual control. To implement this, we must have both the first and second functions. If the actual material and kanban can consistently move together, the following become possible:

1. No overproduction will occur
2. Priority in production becomes obvious (when the kanban for one item piles up, that is the item that must be produced first)
3. Control of actual material becomes easier

The term *kanban* emerged from the standard operations list, which will be discussed later in this book. It came from the practice whereby foremen in separate workplaces would write their group's work content on slips of paper, and post them, along with other foremen's, to show what was going on. In other words, they were putting up their shingles *(kanban)*, so to speak, and hence this term. When one has his kanban up, it is a declaration that his kanban does not lie. "If there is any falsehood contained in what this shop and its kanban stand for, we do not expect any payment." In the old days, merchants declared this to their customers; our kanban lives by that tradition.

The kanban system also clears the air of the erroneous assumptions that outsiders may have of us. It is intended to show that we do things openly.

Six Rules for the Kanban

The better the tools, the more effective they are in attaining the goals. However, if they are used the wrong way, they could turn into weapons which would prevent reaching the very goals for which they were created.

The same can be said about the kanban, which is a tool created to manage the workplace effectively. Here the preconditions for operating kanban, or the rules of the kanban, are explained.

Rule 1: Do Not Send Defective Products to the Subsequent Process

Making defective products means investing materials, equipment and labor in something that cannot be sold. This is the greatest waste of all. It is the worst offense against cost reduction, which is the goal of any industry. If a defective product is discovered, measures to prevent its recurrence must be taken ahead of everything else, to make certain that similar defectives will not be produced again.

To thoroughly implement activities for eliminating defects, the first rule must be that defective items will not be sent to the subsequent process.

Observation of the first rule means the following:

1. The process that has just produced a defective product can immediately discover it.

2. The problem in that process is immediately called to everyone's attention. If it is left unsolved, the subsequent process may stop or the process itself may be saddled with a pile of defects. Thus, managers and supervisors are forced to engage in the task of undertaking measures against recurrence.

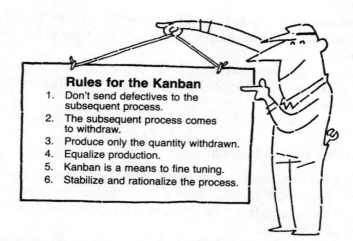

Figure 19. Rules for the Kanban

In order to abide by this rule scrupulously, machines must be made to stop automatically when they produce defectives, and workers must stop their operations. This is where the concept of automation with a human touch comes into play.

If defective products get mixed up with good products, exchange them promptly. If defective products are supplied by subsidiaries, do not modify their delivery cards. Ask them to replace the exact number of defective items in the next delivery.

Unless there is an assurance that parts flowing through all the processes are good products, the kanban system itself will collapse.

Rule 2: The Subsequent Process Comes to Withdraw Only What Is Needed

The second rule is that the subsequent process comes to the preceding process to withdraw parts and materials at the time needed and in the quantity needed.

This aspect has already been discussed in connection with the issue of just-in-time. A loss is created if the preceding process supplies parts and materials to the subsequent process at the time it does not need them or in a quantity above the latter's needs. The loss can come from many quarters, including a loss from excessive overtime, a loss from excess inventory, and a loss from investing in new facilities without knowing that the existing facility is actually sufficient. Then there is a loss arising from the inability to take countermeasures when the existing facilities create a terrible bottleneck, again without knowing the exact situation. The worst loss arises when the process cannot produce what is necessary, because it has been producing what is not necessary.

To eliminate these types of waste, the second rule comes into play. To observe this rule, what must be done?

If we abide by the first rule that no defective products be shipped to the subsequent process, the process in question can always discover defects appearing within the process. There is no need to obtain information from other sources. The process can supply good quality parts and materials. However, this process does not have the ability to determine the time and quantity the subsequent process will require of its products. In order for the process to function properly, this information must be supplied by another source.

Therefore, we have changed the thinking from "supplying to the subsequent process" to one of "the subsequent process coming to withdraw" from the preceding process at the time needed and in the quantity needed.

From the final assembly line, which is the final process, to the process where materials leave the storage room, which is the first process, if all processes can agree on a procedure whereby the subsequent process goes to the preceding process to withdraw materials needed at the time needed and in the quantity needed, then no process has to worry about information concerning the time and quantity of the materials to be supplied to the subsequent process.

By transforming the notion of "supplying" to one of "withdrawing," we have discovered in one stroke the means of solving a very difficult problem. This is the foundation of the second rule, that the subsequent process must come to the preceding process to withdraw. But a number of concrete steps are needed to ensure that the subsequent process will not arbitrarily withdraw form the preceding process. They are as follows:

1. No withdrawal without a kanban
2. Items withdrawn cannot exceed the number of kanban submitted
3. A kanban must always accompany each item

These three major principles ensure that the second rule will be correctly carried out.

Rule 3: Produce Only the Exact Quantity Withdrawn by the Subsequent Process

The importance of the third rule, to produce only the exact quantity withdrawn by the subsequent process, can be inferred from the discussion of the second rule. It is, after all, a logical extention of the second.

Of course, this rule is predicated on the condition that the process itself must restrict its inventory to the absolute minimum. For this reason, the following must be observed:

1. Do not produce more than the number of kanban
2. Produce in the sequence in which the kanban are received

Only through observance of these operational guidelines will the third rule become functional.

One further consideration is that by observing the second and third rules, the entire production process can function in unison, almost like a single conveyor. Simultaneity in production can be accomplished.

We all know how the introduction of the conveyor line has contributed to the standardization of operations and cost reduction. With that in mind, we can appreciate the significance that simultaneous production holds in our overall production schemes.

Rule 4: Equalize Production

In order to observe the third rule, to produce only the exact quantity withdrawn by the subsequent process, it becomes necessary for all processes to maintain equipment and workers in such a way that materials can be produced at the time needed and in the quantity needed. In this case, if the subsequent process comes to withdraw materials unevenly with regard to time and quantity, the preceding process will require excess personnel and facilities to accommodate its requests. The end result is that the earlier the process stands in the total manufacturing process, the greater the need for excess capacity.

Needless to say, this is something that absolutely cannot be tolerated. Yet, if the preceding process has no excess capacity at all, it may not be able to deal with the requirements of the subsequent process without resorting to producing materials ahead of time when it has time on hand. This is a clear violation of the third rule; and, of course, we do not allow any violations of the rules.

This is where the fourth rule, which insists on load smoothing (equalizing) in production, comes in. As discussed earlier, the load-smoothing system of production is the very foundation of the Toyota system.

Rule 5: Kanban Is a Means to Fine Tuning

One of the functions of the kanban has been described as an automatic directional device containing information for workers concerning their work order.

Therefore, when the kanban system is adopted, we can dispense with the start-up plan chart and the transportation plan chart which are normally provided for the workplace. For the workers, the kanban becomes the source of information for production and transportation. Because the workers must rely heavily on the kanban to do their work, the load-smoothing system of production becomes extremely important.

What kind of problem can arise if there is no load-smoothing system of production?

Assume that a certain metal-stamped part takes four hours to produce, from the time the die is set to the time the part is stamped and sent to the next process. Assume, further, that the kanban is organized so that when inventory of this particular stamped part falls below five hours, an instruction is given to commence the die setup procedure.

However, the subsequent process has increased its production by 100 percent, and the inventory which is supposed to last five hours is withdrawn by the subsequent process in two and a half hours. Now the metal-stamping process has no parts available for an hour and a half (4 hours − 2 hours 30 minutes = 1 hour 30 minutes).

Does this mean that the process in question must create an inventory to last ten hours, which is twice as much as it normally has? The answer is no; when the amount of production is normal, it will be saddled with an excess inventory. Yet, it is not allowed to speculate whether or not the subsequent process may withdraw more next time. Nor is the subsequent process allowed to come to the preceding process and ask, "Can you start the next lot a little earlier, please?" No one is permitted to send information other than that which is contained on the kanban. We do not want confusion in the workplace. In using the kanban system, it is important to abide by the principle of load smoothing in production.

Please review this example to obtain its full message. As it demonstrates, kanban can only respond to the need for fine tuning, but not to a major change. Kanban's full potential is realized when it is used effectively for fine tuning.

Rule 6: Stabilize and Rationalize the Process

The fourth rule, which requires load smoothing in production, is effective in guaranteeing an adequate supply for the subsequent process and, at the same time, in fulfilling the objective of producing materials as inexpensively as possible. In following this rule, we must not forget the sixth rule, which requires that the process be stabilized and rationalized.

In studying the first rule, to refrain from sending defective products to the subsequent process, we have learned the importance of automation with a human touch. If we extend the meaning of *defectives* beyond defective parts to include defective work, then the sixth rule becomes easy to understand. Defective work exists because there is not sufficient standardization and rationalization of work. When waste, unevenness and unreasonableness *(muda, mura* and *muri)* exist in work methods and work hours, they can result in the production of defective parts. Without resolving this issue, no guarantee can be given to the subsequent process that there will be an adequate supply, or that the products can be inexpensively produced. Efforts toward standardization and rationalization of the process are key to the successful implementation of automation. The load-smoothing system of production requires this kind of support to become truly effective.

A lot of effort is necessary to observe these six rules. If the kanban system is introduced without them, it cannot function effectively. As long as you recognize the utility of kanban, you must be prepared to observe all the rules regardless of difficulty. That is the only way to bring about a true cost reduction for your company.

Circulating the Kanban

Here are some concrete steps for circulating the kanban.

A sub-assembly line makes products named A, B, C and D. The parts needed for these products are called *a, b, c* and *d*. The sub-assembly line and the processing lines making parts are separated. There are three processing lines, of which one produces parts *a* and *b*. This is illustrated in Figure 20.

A few pieces each of parts *a* and *b*, made by processing line 1, are stored behind the line with the kanban from that processing line attached to each piece.

The sub-assembly line assembles A. It goes to processing line 1 to withdraw part *a*, and for this purpose it must take the sub-assembly kanban (called *withdrawal kanban*).

It goes to store **a**, withdraws boxes in the required quantity and removes those kanban (called *in-process kanban* or *production-ordering kanban*) that are attached to the boxes. It then attaches the withdrawal kanban which it brought along to receive the parts.

Thus, at store **a** of processing line 1, there are as many detached in-process kanban as there are boxes withdrawn by the sub-assembly line, its subsequent process. All processing line 1 has to do is count the number of in-process kanban, produce the exact quantity and replace those parts in store **a**.

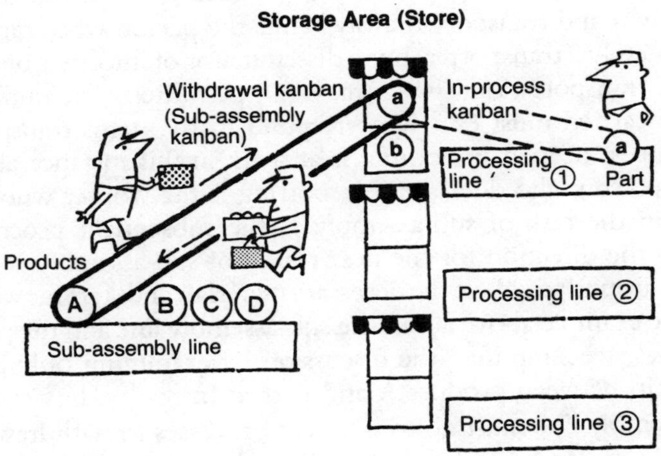

Figure 20. How to Circulate the Kanban

In this manner, all processes are connected by the kanban as if they were one big chain.

Now, suppose trouble occurs, and part *a* is not available in store **a**

In such a case, the subsequent process simply hands its withdrawal kanban (in this instance, the sub-assembly kanban) to processing line 1. At that juncture, processing line 1 interrupts all of its other work and devotes itself fully to the manufacturing of part *a* for which a shortage has been noted. As soon as these parts are made, they are delivered to the subsequent process (in this case, the sub-assembly line).

A Whirligig Beetle

Do you know an insect called a whirligig beetle? It slides over the surface of the water and then suddenly changes its course. It moves with great swiftness. There is something very similar to the whirligig beetle that runs around Toyota's factory.

Under the kanban system, in which the subsequent process must withdraw parts it needs, frequent transportation of parts becomes necessary. Our whirligig is a little train which transports the parts on push carts, each of which is the size of a TV dinner tray.

Withdrawal by the subsequent process means that directions are given to the preceding process to produce those items that are withdrawn and transported away. Thus the person who transports is not merely a transporter, but a disseminator of information. Half-hearted transportation, therefore, is not permitted. The number of kanban carried must exactly correspond to the items transported. The process of load smoothing is at work in this instance also.

The person who does the transporting is the worker who is engaged in the task of sub-assembly at the subsequent process. As soon as the direction for the next round of sub-assembly is given, he goes around to collect the necessary materials and start his work.

In the example above about the sub-assembly line and the processing lines, we simplified the discussion by examining only the relationship between product A and part *a*. In reality, however, the sub-assembly line must go to different processes to withdraw parts needed for product A and place them on push carts all mixed together.

The Storage Area Called Store

We call the place where we hold things a *store*. As you may have guessed, our just-in-time system is sometimes called the supermarket system. We call our storage area a store because it is the place that awaits the coming of customers (the subsequent process).

One of the rules of the kanban system is that no defective item can be sent to the subsequent process. This is the same as the resolve of a store owner not to sell to his customer any defective product.

The space assigned to this store is the space that is needed, since the flow of most products and parts is already known through the load-smoothing system of production. For chassis, for example, the storage space is for five units, and for headlamps the space is for ten units. That is all that is needed.

When stores are set up and only a few items are placed in each store, the workplace becomes easy to observe. We can implement visual control effectively.

If there are too many things in each store, it may mean that capacity is raised excessively. If each store is practically empty, it means that everyone is overworked. All of this becomes clear to everyone at a glance.

The store is clearly defined by a yellow line or by a room divider shelf. Nothing is allowed to be stored beyond these boundaries. In this way, overproduction and over-progression are also discouraged.

Full-Work System

The kanban system has given us an opportunity to provide frequent directions, along with the ability to prevent the flow of too much information. The primary purpose of the kanban system, after all, is to restrain overproduction (as we say, "Do not produce more than is withdrawn").

But in the automated machining processes, there are occasional problems.

Each of the machines is supposed to produce in accordance with its own capability. But no two machines are alike. Imbalances may be created among machines; some machines may produce more, or there may be a breakdown of a machine in the subsequent process. But the machines in the preceding process continue to produce without regard to what is happening in the next process.

What measures can companies take to prevent this from happening if no workers are overseeing these automated machines? How can we let the machines learn how to produce only the quantity required by the subsequent process? To do so, we have invented the so-called full-work system.

Even though the subsequent process may have "too much to eat and be about to vomit (full work)," the preceding process may continue to produce. To avoid this, we have installed limit switches to restrain machines in the preceding process from processing continuously. The action of the limit switches is automatic.

For example, the standard stock on hand for a certain machine is established at five units. If there are only three units, the preceding process automatically begins to process and continues until the quantity reaches five units. Once this predetermined quantity is reached, the preceding processes stop automatically one after another. Thus, there can be no further unnecessary processing.

This works the same way with the subsequent process. If the standard for stock on hand for the subsequent process is set at four units, and it is one short, the current process immediately begins processing and sends the finished product to the subsequent process. Once the subsequent process has the required amount of stock, the current process stops processing.

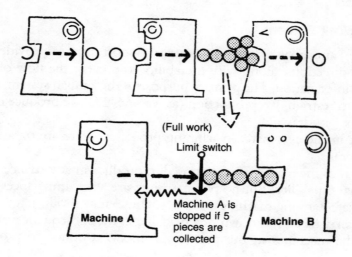

Figure 21. The Full-Work System

All processes are interconnected in this manner to ensure that each will have the exact quantity of standard stock on hand. As shown in Figure 21, wasteful processing is always avoided.

When processes are separated and each has workers, the kanban performs the function of the limit switch. Since this full-work system performs the function of the kanban in fully automated processes, sometimes it is known as the *electric kanban*.

Uses in Unexpected Areas

To prevent unevenness in rotation, balancing weights are attached to a propeller shaft. There are five types of balancing weights. We choose some appropriate ones depending on the degree of unevenness developed in the propeller shaft. Of course, if there is no unevenness, no balancing weight is required. But on occasion, several pieces must be attached. The use of these five types of balancing weights thus varies greatly and is irregular. Unlike most other parts, whose quantities are known through the production plan, the amount of use of balancing weights cannot be predicted.

With parts like this, unless control is adequate, there may be a demand for a special order, on the one hand, and unused parts piling up in storage, on the other. Thus, the start-up and transportation plans must be frequently revised. The result is that we encounter serious difficulties of waste and unevenness in all the processes related to the balancing weights — from production and transportation to their use. In fact, in spite of all the efforts put forth, we gave up thinking that in the case of parts like this, we could do nothing. That is, until we introduced the kanban system.

To control all the processes, we started by always having the exact information concerning the parts (the five types of balancing weights) in storage. Next, so as not to create an emergency need or an oversupply, we made our starting and transporting procedures reflect the actual conditions of storage, and introduced the kanban system to assist in the process.

As a result, we have been able to solve all the problems and keep the exact information concerning the parts. In other words:

• By attaching a kanban to the actual item, it can always be accurately recognized.

- With the kanban making rounds between the processes, starting and transporting procedures can be conducted at all times in the right sequence.
- These have resulted in our ability to maintain required quantities for the five types of balancing weights at a constant level, enabling us to reduce the amount in storage significantly.

This is an important example. Often people say that kanban can be used only for controlling parts which are used every day on a constant basis. It is true that in the rules governing the kanban system, some of the important conditions include the stabilization and equalization of products. But it does not mean that kanban cannot be used without stability in the quantity of parts withdrawn.

Kanban should never be regarded merely as a tool of control for commonly shared parts and universally used parts for which user quantities are constant. It is just as effective in controlling specialized parts not in constant use, which at first glance may appear not to be receptive to the control of kanban.

Generally, in producing parts whose quantity of use is irregular or not stable, it is important to remember not to delay the relaying of information. Since the quantities are irregular and inconstant to begin with, if information is delayed the existing distortion becomes even worse. It is essential that kanban be circulated frequently to solve this problem.

In contrast to parts whose production quantities are stable, those which are irregular or inconstant may at times require you to have more materials or parts on hand. This much is admittedly true.

Even more important than eliminating a delay in the transmission of information is eliminating the delay in processing and shortening the lead time. We have already discussed lead time in Chapter 2, so shall not say any more about that subject here.

Fewer Kanban

In implementing the kanban system, care must be taken not to overdo it. In other words, the fewer the number of kanban, the better.

One of the functions of kanban is to transmit information to the preceding process indicating what the current process needs. If there are too many kanban, the information is no longer accurate. For example, many parts are needed for a sub-assembly process, but if there are too many kanban pieces, one does not know which part is needed at the moment.

We have discussed again and again the importance of kanban, making it into a tool that controls and improves the workplace. Reducing the number of kanban has the effect of restraining as much as possible the number of start-ups within the process. In this way, existing problems can be made apparent — a useful function of the kanban. But if there are too many pieces of kanban, they can hide problems. What is the use of kanban, then, if we simply use them to be attached to or detached from parts and materials?

Simply put, the fewer the number of kanban, the better. With fewer, their sensitivity is greater. Beware of the propensity to overdo the kanban. It is a route marked with failure.

Use your mind wisely. There are many uses for kanban to raise the performance level of your workplace. We often say that the level of a given workplace is known by its use of kanban.

So far, we have explained the basic thinking and content of the kanban system. However, as far as the kanban system is concerned, unless one has actually experimented with it, one can never truly understand its essence.

"Improvement is eternal and infinite." And at Toyota we say:"The use of kanban must not be limited to the preservation of the status quo. The task given to the people connected with the kanban system is to use their imagination and effort to let it develop further."

6
Reality of Workplace-Determined Standard Operations

Three Components of Standard Operations

Standard operations serve a number of useful functions. For example, workers can rely on them to raise their productivity, the foreman can use them as the basis for managing his process and improvement activities groups can use them as their foundation. In standard operations, it is necessary to combine people and things as effectively as possible, taking into account the conditions that are required in productivity improvement. In the Toyota system, this process of combining people and things is called a *work combination*. The sum total of all work combinations constitutes standard operations.

There are three components to standard operations. They are:

- Cycle time
- Work procedure (work sequence)
- Standard stock on hand

Without any one of these components, standard operations cannot exist.

One of the characteristics of the Toyota system is that it is the foreman who determines standard operations. In most other companies, such standard operations are determined by an engineer or a technician who, although a third party to the work process, is presumably conversant with the techniques of industrial engineering.

The standards are exactly those determined by the foreman, and he instructs his workers to abide by them. He must be able to demonstrate to his workers at an appropriate speed how to follow the standards. The speed must also seem appropriate to an impartial observer.

The foreman knows well the process in his workplace for which he is responsible. When he determines standard operations, naturally he is expected to be able to follow them. In determining the standards, he must be willing to assume the role of a teacher to workers who are his apprentices. He must investigate thoroughly the bad conditions surrounding standard operations, such as difficulties in implementing them.

This approach may appear unscientific. But the foreman determining the standards is the one with a record of accomplishments and knowledge. He can determine the standards with confidence — and it is a sound method.

Once established, standard operations must not be left alone in a complacent manner. The standards are living things which are never completed. They have the built-in task of remaking themselves continuously. The foreman who determines them and the workers who abide by them must always be mindful of the need to improve them. Based on that need, revision continues.

If the standard operations sheet remains the same for long, it is proof positive of the incompetence of the foreman.

How to Determine the Cycle Time

The term *cycle time* refers to the time frame of so many minutes and so many seconds needed to produce a unit or a piece of product. This is determined by the amount of production and the operating time.

To obtain the cycle time, first divide the quantity required for a month by the number of days of operation to produce the required quantity per day. The daily operating time divided by the required quantity per day becomes the cycle time.

$$\text{Required quantity per day} = \frac{\text{Required quantity per month}}{\text{Number of days of operation}}$$

$$\text{Cycle time} \; = \; \frac{\text{Daily operating time *}}{\text{Required quantity per day (unit)}}$$

Once the cycle time is determined, the amount of work for each worker is determined so that he can do his work within the constraint of the cycle time. No allowance is made for an extra margin, as is often done in operational studies.

The foreman also determines the speed, degree of skill and other standards that are required. When a newcomer becomes competent enough to do his work at the same speed as the foreman, he becomes a full-fledged member of the team.

When the cycle time is determined in this manner, individual differences emerge, depending on the worker. And since no margin is allowed, waste is immediately discovered. This is directly connected to improvement. If a certain work is slightly off the cycle time, the group or individual can strive to correct it to fit the cycle time. This is the beginning of improvement.

Work Procedure (Work Sequence)

The term *work procedure* refers to the order in which work is performed, in accordance with the flow of time. For example, when a worker is engaged in processing, he must transport materials, and attach and remove them from machines, as he observes them turning into products with the flow of time. The term does not refer to the order in which the flow of products is registered.

If the work procedure is not made clear, each worker may perform his task any way he likes. And even the same person may perform the same task differently each time he does it.

If the work procedure is not followed correctly, someone may forget to process, while another may attach a wrong part to the machine and send a wrong item to the subsequent process. The machines may be broken, and the assembly line may be stopped. In the worst-case scenario, a car liable for recall may be produced.

---

* If there are differences in the straight numbers for the daily operating time, calculation is made on a per hour basis.

The work procedure must clearly define everything minutely and quantitatively. Everything must be expressed in concrete terms. This will assist the foreman in avoiding waste, unevenness and unreasonableness when determining standard operations. For example, how both hands are to move, where to place both feet and the general nature of the work must all be clearly stated. Workers must be able to understand the procedure, and it must be standardized. The intent of the person establishing the work procedure must be clearly discernible, as if to say, "This is the way I want you to do your work." With that understanding, workers can abide by the work procedure with confidence to make high-quality products safely and quickly.

Standard Stock on Hand

The term *stock on hand* refers to parts and materials that are essential in starting up work within the process. It includes those parts or materials already attached to the machines.

Changes occur in the standard stock on hand either through the way machines are placed or the manner in which the work procedure is determined. The controlling factor in determining the standard stock is simply that the work cannot progress without so many pieces of parts or materials on hand.

If the machine placement is the same and the operation follows the order established for semi-finishing processes, there is no necessity to have the standard stock on hand between processes except those already attached to the machines. However, if the operation is to progress in the inverse order of the processes, then one piece each of stock on hand is required between each of the two processes. (Or two pieces, if two are to be attached to each machine.)

The standard stock on hand must be increased to meet any one of the following conditions: when additional pieces are needed to perform quality check; when the temperature must fall by so many degrees before the next operation can commence; and when cleaning the machine to remove oil.

One may confuse standard operations and work standards as one and the same, but they are not. *Work standards* refers to those standards that are needed to implement standard operations. For example, in heat treatment, consistent with the quality of material, standards must be set for the degree of heat, time and coolant. And in machining processing, standards must be set for cutters or bits in the following areas: form, configuration, quality of material, size, conditions for cutting, cutting oil, etc. In establishing these standards which are needed to produce adequate quality, economic conditions for operations are also taken into account.

Once determined, standard operations are posted, in the form of a *standard operations bulletin,* in an easily seen location at each workplace. The bulletin becomes a guide for new workers. For older workers who may not need such guidance, the bulletin still serves as a restraining device when they want to engage in work other than standard. When work is performed on the basis of the standard operations bulletin, and if an inconsistent element is found, that discovery can serve as the starting point for improvement. The next step is issuance of new standard operations or guides.

For supervisors and managers, the posted guidelines and standard operations bulletins show at a glance if the workers are doing their work correctly, or if points in the guidelines and bulletins require revision.

Reproduced on the following pages are some standard forms in use at various headquarters factories of Toyota. Forms may vary slightly from factory to factory, but they share basic features. When adopting these forms for use in your own company or factory, simply choose the ones that best suit your purpose.

Methods of Determining Standard Operations

1. Table of Part-Production Capacity

To determine standard operations, first enter the production capacity for each part in the table of part-production capacity. This must be done by each process.

The following must be entered in the table: work sequence, name of the process, machine number, basic time, tool exchange time, units and production capacity.

This table is an important one, because in determining standard operations, it becomes the basis for establishing the work routine. On the following page, an example of this table is given.

Revised, February 5, 1975   Pg. ___ of ___ pages

Section chief	Floor foreman	Table of Part Production Capacity	Item no. 43202 – 36022		Type		Group's name	Worker's name
			Item name		No. of units			

							RU	JU	HU
							2	2	2

Work sequence	Name of the process	Machine number	Basic time — Manual operation time (min. sec.)	Basic time — Machine processing time (min. sec.)	Basic time — Completion time (min. sec.)	Tool exchange — Exchange unit	Tool exchange — Exchange time	Production capacity (960') units	Notes
1	Shoving and pushing both centers	CE–239	08		1 18	140	1'00"	630	Center drill
2	Outside diameter, rough planing	LA–1306	08	1 27	1 35	10	30"	530	Tip bite (in feed)
						20	30"		Tip bite (in feed)
						80	30"		Tip bite (outside diameter)
3	Outside diameter, semi-finishing planing	LA–1307	08	1 24	1 32	10	30"	548	Tip bite (in feed)
						20	30"		Tip bite (in feed)
4	Outside diameter, finishing planing	LA–1101	10	1 32	1 42	40	30"	488	Tip bite (in feed)
						20	30"		Tip bite (outside diameter)
2-1	30 φ outside diameter grinding	GR – 120	(15)	(2 21)	(2 30)	1,500	70'00"	} 680	$\frac{9"\;6"}{2} = 2'21"$
2-2	30 φ outside diameter grinding	GR – 121	(12)	(2 21)	(2 27)	1,500	70'00"		$\frac{6"6"}{2} = 2'21"$
	(Two stands of machines for the same process)		14						Manual operation time per unit $\left(\dfrac{15"+12"}{2} = 13.5" \rightarrow 14"\right)$
3	Broach removal (two units)	BM – 131	(09)	(43)	(52)	700	5'00"	1,937	
	(Two units or more processing at a time)		05		26				Manual operation time per unit $\left(\dfrac{9"}{2} = 4.5" \rightarrow 5"\right)$
4	30 φ measurement (1/5)		(20)						
	(Measuring one unit in every five units)		04						Manual operation time per unit $\left(\dfrac{20"}{5} = 4"\right)$
	Total								

manual operations ——— machine processing - - - - - -

Figure 22. Table of Part-Production Capacity

2. Standard Operations Routine Sheet

After the production capacity for each part is noted in the above table, obtain the cycle time from the required quantity per day and hours of operation. Next, determine in what sequence each worker will work within that time frame. If it is a simple one, the work sequence can be derived directly from the table of part-production capacity. If it is more complex, one may not know if the machine has already completed its automatic machine processing while the work sequence is still being prepared.

The *standard operations routine sheet* is a tool designed to show the passage of time at a glance to help in the determination of work sequence. It contains information on work sequence, work content and operations time (work hours).

The column for noting the operations time is graduated in a unit of seconds. Normally, one sheet is adequate for entering operations lasting two minutes (or in some cases, three minutes). When the operations time exceeds two minutes, or when there are different types of operations, vertical and horizontal lines can be added to let one sheet cover all the required information.

The foreman must follow the standard operations routine sheet and do the work himself. He must ascertain if he can do a good job in the sequence and within the cycle time given.

Once he can do the job well in accordance with the standard operations routine sheet, he must teach those operations to his workers until everything is fully understood.

3. Operation Pointers

This document is given to the workers, showing them what to look for when engaged in a specific operation. The work sequence is determined and written down for each of the following processes: operating the machine, exchanging cutting tools, exchanging and setting up dies, processing parts and sub-assembling.

The work content is noted in accordance with the work sequence, and the most critical points in performing the task are explained. To make it easier to understand, detailed information is given along with a drawing for each item. Abstract expressions are avoided. The description is one of concreteness, and the quantity for each item is clearly specified.

Item no. Name of item	4320 ½—36022	Standard operations routine sheet No. 1		Date of manufacturing	Feb. 5, 1975	Required daily quantity	255	Manual operation ———
Process	Steering knuckle machining process			Worker's group		480 minutes/ required quantity (cycle time)	1 minute 53 seconds	Machine processing – – – Walking 〜〜〜

Work sequence		Name of operation	Time Manual	Machine	Operations time (unit: one second)
1		Removing rough materials from the pallet	01"	—	
2	CE–239	Work removing and attaching, starting the machine	08"	1'10"	
3	LA–1306	Work removing and attaching, starting the machine	08"	1'27"	
4	LA–1307	Work removing and attaching, starting the machine	08"	1'24"	
5	LA–1101	Work removing and attaching, starting the machine	10"	1'32"	
6	DR–1544	Work removing and attaching, starting the machine	07"	34"	
7	SP–101		07"	1'02"	
8	MM–122		04"	—	
9	HP–657		10"	17"	
10	BR–410	Work removing and attaching, starting the machine and wash	13"	54"	
11		Attach the nipple, put completed work in the pallet	15"	—	

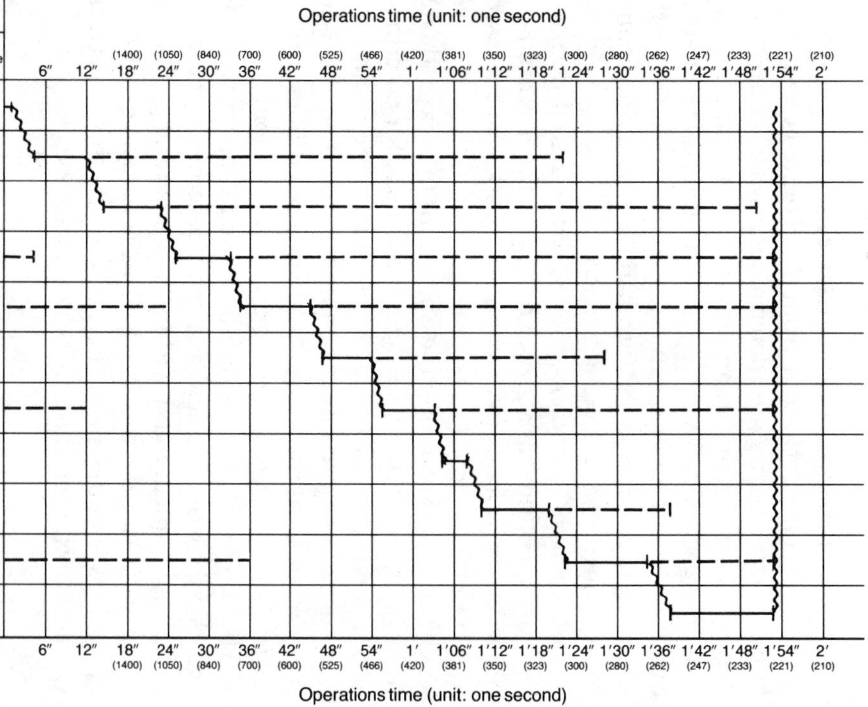

Operations time (unit: one second)

6" 12" 18"(1400) 24"(1050) 30"(840) 36"(700) 42"(600) 48"(525) 54"(466) 1'(420) 1'06"(381) 1'12"(350) 1'18"(323) 1'24"(300) 1'30"(280) 1'36"(262) 1'42"(247) 1'48"(233) 1'54"(221) 2'(210)

Figure 23. Standard Operations Routine Sheet No. 1

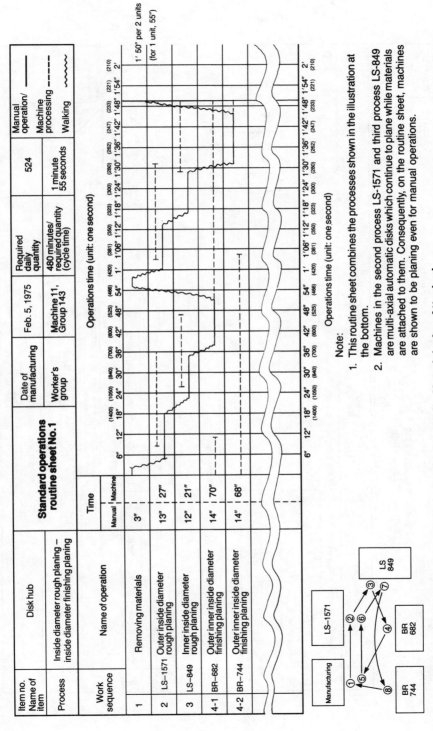

Figure 24. Example of Machines Continuously Planing While Materials Are Attached

Section chief	Floor foreman	Group leader	Operation pointers for cutting tool exchange			Group's name	Worker's name

RS type ring gear processing

Line name			
Work sequence	4	Machine no.	LS-494
Name of process	Inside diameter semi-finishing, chamfering		

Drawing

L-type wrench 4 mm

(A) (B) (C) (D) (E)

Cutting areas

No.	Work content	Critical areas (right and wrong, safety and ease in operation)
1.	Tighten the tip and loosen the bolt	While using your left hand to hold the breaker 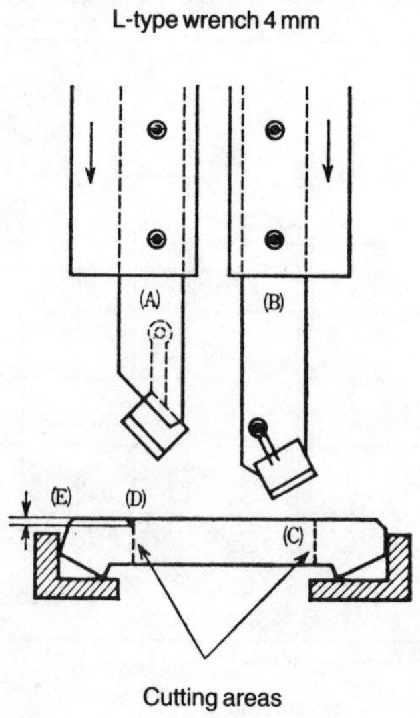
		With the 4mm L-type wrench
2.	Remove the tip	
3.	Clean sides of tip and holder where previously attached	
4.	Attach new tip breaker	All the way in
5.	Tighten bolt to secure tip breaker	While holding the tip tightly
6.	Repeat the same procedure for (B)	The inside diameter of semi-finishing (c)
		Targeted measurement: 107.56-58 (3) chuck
		107.66-68 (9) chuck
		Chamfering of (D) 45° x 2.8c (after completion of tempering and grinding, 2.5c)
		Depth of (E)'s corner (inside diameter chamfering) to 1.8 - 1.9 mm
		Attach a dial to make inside diameter adjustment
		Let the holder move slightly
		Tip exchange (A) 240 pieces
		(B) 240 pieces

Figure 25. Operation Pointers for Cutting Tool Exchange

4. Manual on Work Directions and Standard Operations Bulletin

The *manual of work directions* serves as the basis for giving instruction on performing standard operations accurately.

This manual is written based on the table of part-production capacity and the standard operations routine sheet. It gives the work contents for each person, consistent with the line's amount of production, and explains critical areas for safety and quality in the order of the work sequence. It provides illustrations for machine placement dealing with the work of one person. It also notes the cycle time, work sequence, standard stock on hand and methods of engaging in quality check. In preparing this manual, take care to ensure that if they follow the directions given in this document, workers can work with assurance, speed and safety.

The drawing of machine placement in the manual of work directions is normally noted on a separate sheet of paper, 12x16½ inches in size, to which columns for work sequence, standard stock on hand, cycle time, net operating time, safety and quality check are added. This completed sheet is placed in a case and displayed in the machining processing lines and sub-assembly lines. We call this sheet the *standard operations bulletin*.

Displaying the standard operations bulletin indicates to the workers what their supervisor expects of them in their work.

The supervisor has many subordinates, and it is difficult to remember the work he has given to each person in his section. However, he can confirm if each of his workers is doing an adequate job by simply looking at the standard operations bulletin. He can also probe into the standard operations to determine if additional waste and defects can be found.

Through this bulletin, managers can evaluate the capability of their supervisors and monitor the work of individual workers. When a mistake occurs, managers not always on the scene can use the bulletin to point out the error. As a tool for visual control, the bulletin is very useful.

Work Combination

Previously we discussed that standard operations are the sum total of all combinations between people, things and machines designed to promote the most effective system of production. Here we shall concern ourselves with the combination itself.

Section chief	Floor foreman	Group leader	Manual on Work Directions	Item number	Required quantity	Group's name	Worker's name
				4320½ — 86022	448 units/day		
				Name of item	Break-down number		
				Steering knuckle	1/3		

Diagram (process layout):

Top row (left to right): SP-101 ⑦ → MM-122 ⑧ → HP-657 ⑨ → BR-410 1/10 ⑩ → Cleaning ⑪ ⑫ Stamping serial number → ⑬ Finished products store

Bottom row (right to left): DR-1544 ⑥ ← LA-1101 1/1 ⑤ ← LA-1307 ④ ← LA-1306 ③ ← CE-239 1/50 ② ← ① Materials store

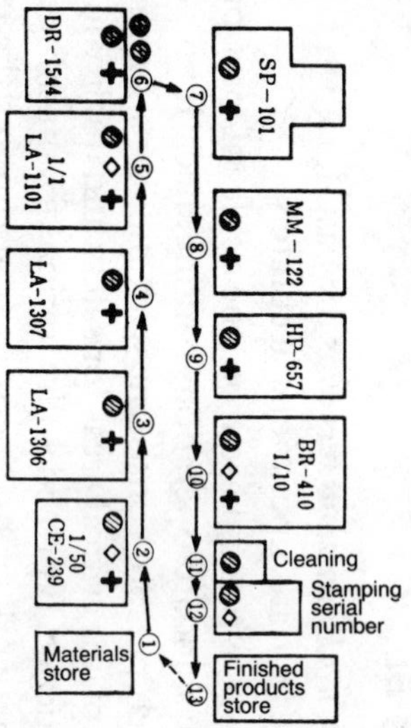

Cycle time	1' 53"
Standard stock on hand	13 pieces
⊘	Standard stock on hand
✛	Safety caution
◇	Quality check
Net operating time	1' 54"

#	Work content	Quality Check	Gauge	Critical Areas (right and wrong, safety and ease in operation)	Net operating time Min.	Sec.
1	Taking out materials			With the right hand		03"
2	CE-239 Detach, attach and start machine	1/50	Visual	If center is too shallow, dangerous for subsequent LA, GR processes		11"
3	LA-1306 Detach, attach and start machine			Both centers must be attached securely		11"
4	LA-1307 Detach, attach and start machine			Both centers must be attached securely		10"
5	LA-1101 Detach, attach and start machine	1/1	C	22.5 +0.25 +0.20 33.1 +0.25 +0.20 Remove scraps by derrick		12"
6	DR-1544 Detach, attach and start machine		Visual	Ascertain penetration from reverse side		09"
7	SP-101 Detach, attach and start machine			Clean scraps from attachment sides on M-22, P-1.5		09"
8	MM-122 Detach, attach and start machine					05"
9	HP-657 Detach, attach and start machine			When bush is let into orifice the top side where oil chamber is cut in the circumference of a circle		12"
10	BR-410 Detach, attach and start machine	1/10		Attachment sides must be cleaned		
		1/10	PS	Brightness over 80%		15"
		1/10	LF	+0.25 −0		
11	Cleaning machine, remove, attach and start					
12	Attach the nipple			Tighten with the impact		17"
13	Stamping serial number, detach, attach and start machine	1/1	Visual	Half full is not acceptable		
14	Finished products stand					
				Time total	1'	54"

Figure 26. Manual on Work Directions: Processing

Section chief	Floor foreman	Group leader	Manual on Work Directions	Item number	45200 — ★★★★★	Required quantity	600 units/day
				Name of item	Steering-post assembly	Break-down number	2/3

Group's name   Worker's name

Sub-assembly stand
CO—129

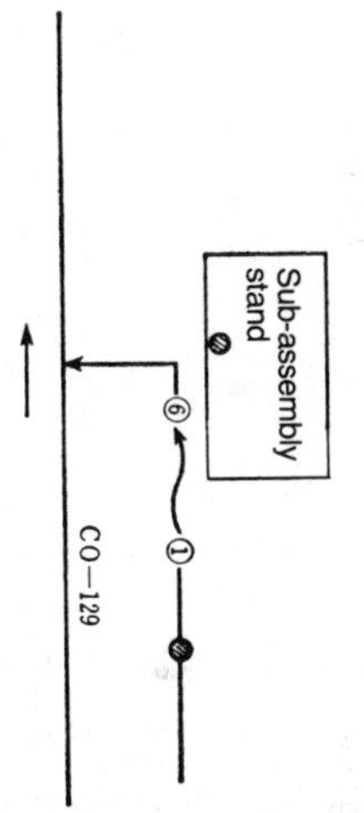

	Cycle time
	1' 24"
	Standard stock on hand
	2 pieces
	Standard stock on hand
	Safety caution
	Quality check
	Net operating time
	1' 22"

Work content	Quality Check	Gauge	Critical Areas (right and wrong, safety and ease in operation)	Net operating time Min.	Sec.
1 Attach gear box to assembly jig			Push in horizontally		03"
2 Place plate thimble on the sector shaft and put in the gear box			While turning the main shaft, put in with the roller of the sector shaft in the middle		15"
3 Attach the sector plate			Bolt tightening torque 600-700 kg/cm		33"
Attach thrust screw					
4 Set the thrust screw			After tightening, return 1/3 to 1/4 to lock		15"
5 Put mask jacket and tighten the clamp bolt			Bolt tightening torque 600-700 kg/cm		12"
6 Detach and hang on the roller conveyor					04"
			Time total	1"	22"

Figure 27. Manual on Work Directions: Sub-Assembly

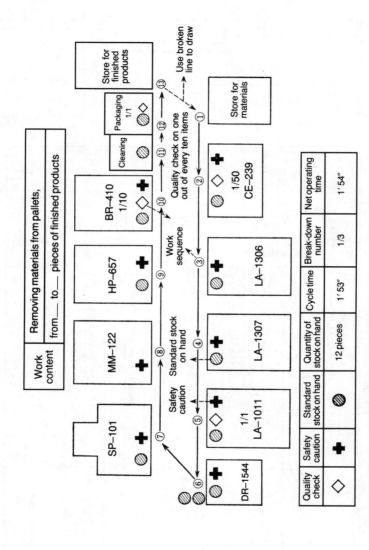

Figure 28. Standard Operations Bulletin

To create an atmosphere conducive to teamwork, the work combination requires that the work area for each worker not be fixed and separate. It should overlap with the areas of responsibility of other workers, making it easier to promote cooperation between different processes.

If an individual's work area is fixed, the person who works faster will progress more quickly, producing a mountain of items ahead of those who work slower. He may enjoy a little more time on hand, while those who are slower may try to catch up and send defectives to the subsequent process. Normally in this situation, the quantity of output is determined by the pace of the slower worker.

To prevent this from happening, the area of responsibility of each worker must be made to overlap with those of others, and his work must be made easier to handle. It should not be conducted like a swimming relay, but like a running relay on the ground.

To make the work easier, the distance between workers (and their machines) must be shortened. Make it easier for them to communicate with each other. In a machine placement combination, avoid an arrangment that only takes into account the capability of one worker to handle one machine.

Teamwork cannot be effected if machines encircle people, as cages confine animals in a zoo. Nor can a man function smoothly if he is treated like a bird in an aviary.

When work areas are allowed to overlap, the momentary absence of a worker can be covered by the two workers on either side. In some instances, absences for an entire day can also be covered. The quantity of output per hour may decrease, but the foreman can extend the work hours to obtain the amount required. This method is especially effective in an assembly line that requires a substantial amount of manual labor.

SAYINGS OF OHNO

Mutual Help (Relays in Swimming and Running)

In a swimming relay, faster and slower swimmers must each cover a set distance. In the case of a running relay, however, the faster person can cover for a slower person in the baton touch zone.

Our line work is like a running relay game on the ground. Supervisors must create baton touch zones in their lines to raise the efficiency of these lines.

Remote Islands

Placing workers here and there means they can hardly help each other. In allocating work and arranging workers, be sure to place them in such a way that they can help each other and combine their work. This can also lead to a reduction in our manpower requirement.

Work Flowing and Work Floating

The process moves forward as things flow. That is what we call the "work flowing." If a conveyor belt is used to transport things, and no further improvement is made, then work is only "floating," not "flowing." Work that floats creates many remote islands and does not allow us to utilize our time on hand effectively.

Overall Efficiency and Balance Between Workers

It is important to produce exactly what is indicated by the cycle time (tact), and no more. If a worker is only concerned with his own high performance, he may create a stockpile of things in front of the worker in the subsequent process. He creates tension among his co-workers and also adds a burden on the company to have someone handle his excess products. For the line as a whole, its own efficiency actually declines.

Generally, where a number of people take turns working, there is always a process that creates a bottleneck, and this is what determines the capacity of the line. If the worker who is responsible for the bottleneck can receive help, the bottleneck will disappear. But in reality, workers who work fast do not want to help their co-workers. Instead, they continue to produce more and create a stockpile of things in front of the bottlenecked process. Everyone's feet will be dragging because of this and the overall efficiency will suffer.

To prevent this, the foreman must insist on his workers observing the standard stock on hand, cycle time and work sequence as determined by the standard operations. When time on hand is created, he can reorganize his work crew to restore balance.

When attempting to balance uneven loads among workers, it is not easy to find a perfect solution. In most lines, especially when they are relatively small with only a few workers, a small imbalance seems to be present at all times.

In such an instance, if the person who has completed work assists those who are slower than he, the efficiency level of the entire line will be raised. When making work assignments, it is good to keep in mind that work boundaries must be drawn in such a way that mutual assistance can be rendered.

There are companies that place a great deal of emphasis on the ability of their individual workers. Yet somehow their factories do not have good performance records. They are slow to dissolve imbalances among their workers. If a mountain of unnecessary inventory is created by their excellent workers, instead of having a cost reduction, there will be a cost increase.

Unfortunately, mistakes like this are common. They happen because companies are reluctant to address the issue of imbalance squarely; they somehow erroneously perceive productivity only in terms of the speed with which an individual can perform.

The performance of an individual worker and the performance of the line have a significant impact on the company's cost picture. Yet priority must be assigned to raising the performance level of the entire factory. The former cannot stand alone without the latter.

How to Implement Standard Operations

Supervisors must drill into the minds of workers that they must strictly abide by the standard operations.

No matter how good standard operations may be, without observance there can be no stability in the processes. When standards are not observed, supervisors must take countermeasures against accidents and defectives. All sorts of waste are likely to be created.

In order to make the workers understand and observe standard operations, the supervisor himself must master them. He must be able to drill into workers the importance of the standards, and show them concrete examples of what their non-observance would lead to. Some pep talks are helpful for promoting in workers the desire to excel and to assume responsibility for quality. When standard operations are not observed, the supervisor must investigate the reason, and amend them into standards that can be followed easily by everyone.

The supervisor must check the results of standard operations implementation. If abnormalities are found, he must investigate thoroughly to determine the cause, and take appropriate corrective actions. When the standards themselves are found wanting, it is his responsibility to correct them and inform everyone concerned about the reasons for his action and the contents of the adjustments made.

It is important for the supervisor to adopt an attitude of always thinking and saying things that are based on facts. He must wander around the workplace constantly to check if the workers are following standard operations, and if they know the key points of the standard operations. He must know the actual conditions prevailing in the workplace and be able to guide workers on the spot about the manner in which their work must be carried out.

Standard operations are the mother of improvement. We can never say that the standard operations we now have are the best, and there is no room for further improvement. Standard operations are the result of improvement after improvement, and they are not static or unchanging. Consider for a moment that the present standards are full of waste, and begin immediately to make improvements.

The world does not stand still. New methods are created one after another. Thus, if trends are simply followed, in effect one is maintaining the status quo. A workplace that has a set of standard operations which never change is, relatively speaking, experiencing a regression, because it is satisfied with the status quo. It does not think that there is really a problem and feels that all the improvement has already been attained. A forward-looking supervisor must rescue his company from this lethargy, and continuously amend its standard operations.

Changes in Work Combination

Technology renews without ceasing, and Toyota's manufacturing technology is no exception. What observers can see in our machine shops and metal-stamping factories today shows little trace of what was once there regarding our machine placement, the way we handle things and our work combinations.

Today we can have a minimum of parts on hand and have everything flow smoothly to the final assembly line. This has become possible because we have placed our machines to follow the sequence of processes differentiated by parts, we have thoroughly imbued ourselves with the kanban system of thinking and we have piled up one improvement after another in our work combinations.

Let us look back to the conditions existing in our machine shops in the early days of the company.

- Each machine was placed independently. At each of the machines, there was always one, and sometimes two, workers.
- While the machine was cutting, the worker stood in front to "watch over" it.
- Parts were put on the floor or placed in a box. The storage area was distant from the machine and parts were often put in a location difficult to reach.
- Roller conveyors were used merely as another storage area, with parts piled on them.
- The work surface of each machine differed from those of others. Some were placed higher and some lower, with no set pattern.
- An inspector inspected the finished products, which were then put in a finished product warehouse before shipment to the assembly line.
- If the quantity of finished products on hand was low, workers were considered idle. A commonly accepted notion was the larger the finished product inventory, the better.

The above conditions were found not only in our machine shops but also in other factories. Here we should retrace the steps we took to change the way we do things. The discussion centers around our machine placement practices.

1. Single Placement — One Person, One Machine

This is the simplest form of placement. For each machine there was one worker. The worker placed a part to be processed on the machine and started it. While the machine was cutting, he either stood by quietly observing it or used a brush to oil the machine or remove the shavings.

This created the waste arising from time on hand. When the machine was cutting, it was the machine and not the worker who was working.

In those days, the time spent standing idly by was measured as part of the standard time, and as such was included in the time spent for processing the part.

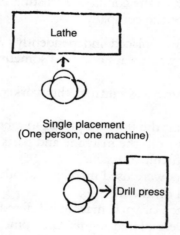

Single placement
(One person, one machine)

Figure 29. One Worker, One Machine

2. Placement By Machine Type — One Worker, Two Machines

The single placement as discussed above had a lot of waste in time on hand and in other areas. To eliminate some of that waste, we thought that we could let the same worker work on another machine by setting up and removing parts there while the first machine did its cutting. We placed the two machines either parallel to each other or in an L-shape and let one worker handle two machines. (This was around 1946-1947.)

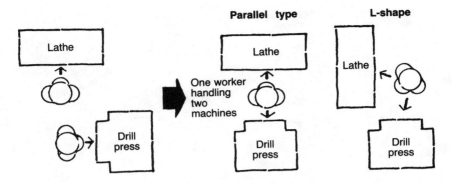

Figure 30. One Worker, Two Machines

This method was a considerable improvement over the system in which one person handled one machine. However, when a worker was made responsible for two machines, he always had to worry about how far the processing had gone in the other machine and could not concentrate on the one he was working on. He could not proceed with confidence to the next phase of work under this system.

The next improvement took into account this issue of confidence in moving to the next phase while the machine was cutting.

We installed an automatic device which would stop a piece from being sent to the next machine if the cutting at the first machine reached a certain point. We also made the machines stop automatically. Shavings were removed by a derrick, and devices were installed to supply oil to the cutting process without a worker having to supply it manually. A study was made to standardize cutting tools (the shapes of drill bits and cutters, and the way they cut). In this manner, workers could move about confidently.

If, after handling two machines, there was still time left, a worker could then handle three machines, which were placed in the shape of a square bracket or a triangle (Figure 31). Better yet, four machines were placed in a square or diamond shape, with only one worker attending. (This occurred around 1949-1950).

By letting a worker attend to several machines of the same type, we were able to raise our per capita production. However, there was a tendency to create only partly completed goods. After going through a drilling machine or a lathe, partly completely goods accumulated within the process and piled up. Also, since parts could not

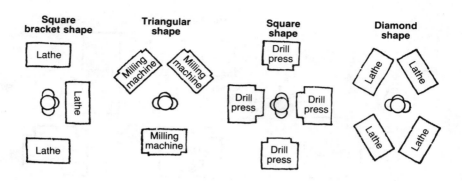

Figure 31. One Worker, Three or Four Machines

be sent to the subsequent process, they took a long time before becoming finished products. To solve this problem, machines were placed in the order of process sequence.

3. Placement By the Process Sequence

With the overproduction of semi-processed goods and an increase in the necessity to transport parts, we discovered that placement by machine type was not the most desirable or efficient arrangement.

Our improvement targets were to restrain the overproduction on semi-processed goods, to transport the manufactured parts as little as possible and to make them into finished products on the spot.

We assembled machines in the order that parts were processed, e.g., a lathe, a milling machine and a drill press. In other words, we shifted gradually from placement by machine type to placement by process sequence (Figure 32).

It was not difficult to find out that placement by process sequence cut to a minimum the distance to be walked by a worker and made it possible to have one worker handle several machines adequately. However, when this was viewed from the overall perspective of the line, the method created several "remote islands." It was not easy to maintain an overall balance. As a result, at each process, goods started to pile up. We could not place workers in accordance with the changes necessitated by the change in number of cars to be produced.

In those days, motion study was frequently undertaken, and machines were placed in such a way that workers could remain

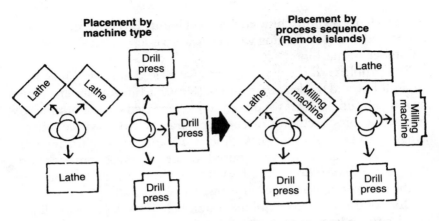

Figure 32. Placement by Machine Type or Process Sequence

stationary. It was then considered desirable for work to be accomplished with a minimum of motion, and walking was considered to be less than desirable. This approach perceived productivity merely as the work efficiency of individual workers. It failed to take into account the synchronized efficiency and method efficiency of the line as a whole.

4. Beginning of the Smooth-Flowing Production System

To make the goods flow smoothly, to raise productivity and to let the workers know that walking was also part of their work, around 1960 we began placing machines in a straight line, freeing

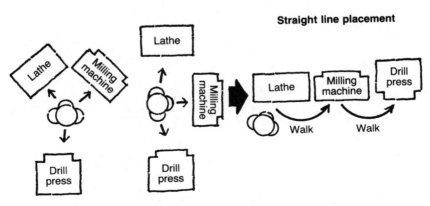

Figure 33. Straight Line Placement

workers from their machine enclosures (Figure 33). This new system had the advantage of letting workers walk while working and handle a number of machines.

But problems did arise. At first, we placed the machines in a straight line and made each group of these machines an independent production line. When we placed workers, based on the number of cars to be produced, we often had a fraction of a whole person assigned to each of these lines. Since we could not assign a fraction of a person, that number was raised to one person. No matter how the group of line workers tried, each tended to overproduce because of the excess manpower, even though it was small.

Our solution was to combine a number of "independent" lines, and let these lines absorb one full person. We were able to make personnel placement based on the changes in the number of cars to be produced. We still follow this work combination and now can produce only what is needed.

7

Improvement Activities for Man-Hour Reduction

Knowing Your Workplace Well

The task of engaging in man-hour reduction activities becomes much easier when the people involved know the conditions of the workplace inside out. As they recognize the waste in their operations, initiate a system of visual control by seeing waste as either time on hand or stock on hand, and use the kanban system to avoid over-production, the man-hour reduction activity may take the following form:

- Eliminate waste
- Redistribute work
- Reduce manpower

Thus, any man-hour reduction activity must begin with the analysis of the conditions existing in the workplace with regard to its operations. Some people, however, may feel that the present system of doing things is just fine. They reason that since the operation rate of the line is satisfactory, and the defect rate is contained within an acceptable limit, the line as a whole performs well. If so, they may become complacent and nip their willingness to work for improvement in the bud.

Workplaces are very much alike. If we observe them closely, their operations can fall within one or more of the categories shown in Figure 34. Examples for each of the categories illustrated are provided below:

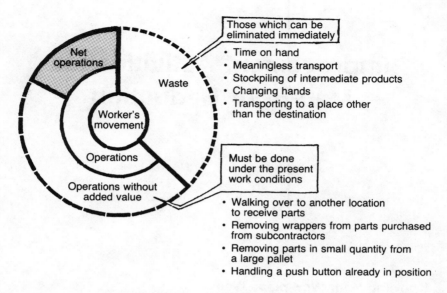

Figure 34. Man-Hour Reduction Activities

Waste (muda): Those operations which are unnecessary for the work and which can be eliminated immediately.

Examples: Time on hand, meaningless transport (stockpiling of intermediate products, transporting to a place other than the destination and changing hands).

Operations without added value: There is no added value to these operations, but under the work conditions prevailing now, these tasks must be performed. (These, incidentally, must be considered as waste.) To eliminate these, the conditions existing in the workplace must be partially changed.

Examples: Walking over to another location to receive parts, removing wrappers from parts purchased from subcontractors, removing parts in small quantities from a large pallet and handling a push button already in position.

Net operations enhancing added value: Those operations which enhance added value can be found in processing (changing the shape, changing the quality or nature of the product, assembling, etc.). In order to manufacture parts and products, work is added to the objects of processing such as materials and semi-processed goods. These operations give added value; the higher their ratio, the higher is their operations efficiency.

Examples: Assembling parts, casting materials, stamping steel plate, welding, tempering the gear, painting the body, etc.

In addition, in the workplace, there are motions other than standard operations, such as adjusting equipment and jigs, and reworking defective products.

When we think in this fashion, we discover that the ratio of net operations that can add value is surprisingly low. Anything other than net operations constitutes factors that raise cost.

The purpose of our man-hour reduction activities is to raise the ratio of net operations as closely as possible to 100 percent.

Redistributing Work

Regarding those operations that have no added value, immediate steps must be taken to eliminate them, if such an action does not cost very much and does not adversely affect the preceding process.

For example, when workers must walk somewhere to receive the parts they need, the time spent in walking may be eliminated by simply providing a shelf close by for parts. After the line is adjusted in this fashion, the foreman redistributes each person's work. The distribution consists of sequentially awarding work containing net operations and other operations that cannot be immediately eliminated within the cycle time.

Figure 35 shows how this redistribution works.

Take a close look at the poor example. Do not distribute equally the excess capacity of worker 4 (which is waste) to four workers, as shown.

An effort is made in this instance to make the waste become apparent. Everyone is aware of the problem and, with this knowledge, of the need to improve. But by averaging out the excessive capacity, the waste becomes hidden. This example awards everyone an excess capacity of 0.15 (waste arising from time on hand). If everyone works under this condition for even ten days, he creates his own slower pace for working. When the next improvement measure is announced, he may become resentful, protesting that the new measure gives him too much added work.

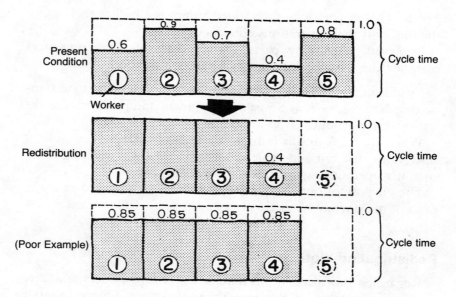

Figure 35. Redistributing Work

Now take a look at the progression from the present condition to redistribution as shown in the illustration. Here, by removing the waste and filling the slot for each one at a time, the work previously done by five people can now be done by 3.4 people. Of course, there is no such entity as a 0.4 person, so one person is needed there. In this example, one person can be reduced immediately, while another person has a work load of only 40 percent.

After eliminating worker 5 through the redistribution, the next important issue to deal with is the fragmentary nature of the work given to worker 4. Addressing the latter first, the next goal is to eliminate this work requiring only 0.4 person.

How can this 0.4 person be removed without difficulty? We must focus our attention on this issue and think through several alternatives. There may be a plan that calls for automating the facility — an expensive undertaking. Or there may be plans stipulating that a chute be provided to reduce the time spent walking to receive parts, or that the size of the pallet be reduced so that needed parts will be handy.

At this stage of the game, do not choose a plan that is overly ambitious. The objective remains to eliminate the work of 0.4 person. Adopt a plan that is consistent with this objective and is inexpensive and easy to implement.

With the completion of these phases of cost-reduction activities, the work which was previously done by five workers is successfully handled by three, reducing the number of workers by two. Now look at the line once more. If you look closely, you may discover waste previously not discovered. You may also find that some operations which you have tolerated — but have never felt comfortable with — do not actually produce added value. Gather up all the information needed, and then meet the challenge of a new task, to eliminate one more person.

This time, proceeding is not going to be as easy as before. Whichever alternative you choose, it means a substantially increased expenditure. And the plans may affect preceding and subsequent processes. It may not be wise to implement any of them at this time. But do not give up. If you observe the workplace every day, knowing what the problem is, one day you may get an idea that allows you to create a fabulous plan.

Your sales may force a change in the cycle time for each of these processes. You may have to modify all the facilities with a model change. When these new needs occur, the plans you have thought about may gain a new luster.

Even if something cannot be implemented immediately, do not give up. Be patient in tackling the problems.

From Work Improvement to Facility Improvement

We have so far discussed man-hour reduction in the following sequence: redistribution of work through elimination of waste, improvement of man hours spent to fill fractional duties and restudy of the entire issue. Returning to the original categories, our process may be rephrased as follows:

1. Eliminate waste immediately
2. For operations that do not produce added value, begin improvement with those that are easy to correct
3. Retain the net operations

In promoting man-hour reduction, there are some problems falling under category 2 above which can be solved by spending money.

In category 3, we have net operations, but it is still possible to reduce manual labor through the introduction of automation. If there is a true need, then automation must be implemented quickly. In other words, all types of operations, whether they fall under 1, 2 or 3, are proper subjects for improvement; in some instances, improvement in all three areas must be undertaken simultaneously and parallel to each other.

Plans for improvement can be divided roughly into two. *Work improvement* refers to changing the work rules, redistributing work, designating where the storage area must be and the like. *Facility improvement* refers to introducing new devices, automating equipment and the like. Again, do not forget that you must start from work improvement and that you must leave no stone unturned in this particular area. Only then can the facility improvement phase begin.

The reasons are as follows:

1. *Facility improvement costs money.* The purpose is to reduce the number of workers in our improvement activities. This can be accomplished equally through work improvement. If a lot of money is spent for facility improvement, it constitutes the wrong approach.

2. *Facility improvement cannot be done over again.* Even when one feels that a certain approach is the best at the planning stage, it can fail. In any endeavor an element of trial and error is always present. If a plan fails, in the case of work improvement, a certain segment can be modified without difficulty. But in the case of facility improvement, all the money invested is wasted.

3. *Facility improvement undertaken at the workplace where work improvement has not been completed is more likely to fail.* Machines are inflexible. Failure is more likely to occur when they are introduced in a workplace without work sequence or work standardization. For example, a metal-stamping machine is automated at a workplace with a poor record in material control. Foreign material is mixed in with the material processed, and the die or the automated device is broken. To prevent this from recurring, a worker is assigned to monitor the automated machine. There is, of course, no man-hour reduction.

These are the reasons why Toyota has insisted on proceeding from work improvement to facility improvement.

This thought process applies equally when automation with a human touch is promoted. Automation, a form of facility improvement, is a means toward man-hour reduction, with the ultimate objective of cost reduction. The trouble occurs with automation when it becomes an end in itself and is introduced without regard to the progress of work improvement in the workplace.

Insufficient work improvement may result in the automated machine — for which a large sum has already been spent — producing defectives within a lot, breaking down frequently and having a poor rate of operation, or requiring a worker to monitor the machine in order to operate it. It is necessary to consider this point very carefully before introducing automation.

SAYINGS OF OHNO

Work improvement means to discover the best method of doing things within the framework of existing facilities. It is not to make equipment. It is to think about the way of doing your work.

People-Centered Way of Thinking

One worker can handle a number of machines when a process is organized to ensure that he can work 100 percent within the cycle time (obtained through the required quantity of output). In this case, it is wrong to consider it a loss if machines remain idle now and then. If an excess capacity exists, it is more effective to seek a lower rate of operation for these machines.

To manufacture beyond the required quantity of output is to create waste. Thus it is better to focus on man instead of machine in organizing work combinations and determining standard operations. This can lead to cost reduction more effectively.

Man-hour is often thought of in terms of the number of people. But while in calculating we can have 0.1 or 0.5 person, in fact the work requiring 0.1 person still needs one full person. Therefore, reducing the work assigned to one person by 0.9 person does not result in cost reduction. It can come about only when the number of real persons is reduced.

Man-hour reduction activities must always focus on reducing the number of workers. When an automatic device is introduced, it may result in a labor saving of 0.9 person. But as long as there is 0.1 person (usually the worker who must monitor the machine) remaining, it is quite possible that reduction in manpower does not occur while, at the same time, a lot of money has been spent. This is still called labor saving by some. At Toyota we call a system "people saving" (*shojinka*) when it can truly contribute to cost reduction, to differentiate it from mere "labor saving."

From People Saving to Fewer People

With the oil shock of 1973, car manufacturers experienced the limit of their expansion. When that happened, it was discovered that reduction of workers through automation did not correspond to the reduction in the number of cars produced. A widely-held view supported a notion that automation meant continuing with a fixed number of workers.

To automate means, in some instances, to transfer or transport goods without the intervention of human labor. For this reason, automated machines became larger and larger.

Workers became performers of auxiliary tasks which the machines could not automate. Thus people began to surround automated machines.

There was no correlation between the number of cars manufactured and the number of people involved in their production. For example, irrespective of the number of cars produced, a certain machine always required three workers to attend to it. This was the reason for saying that automation meant having a fixed number of workers.

But this situation created problems. If the number of cars produced was reduced, then the number of workers involved should also be reduced. From this we began to think in terms of fewer people for the work as well.

SAYINGS OF OHNO

People Saving

You have a labor saving program and you reduce man-hours by 0.5 worker. Yet in reality you have accomplished nothing.

Only when you reduce the number of actual workers involved can you have a viable program of cost reduction. We must proceed to people saving and not labor saving.

Fewer People

You have automated your plant in order to achieve people saving. So far so good, but when reduction in output becomes necessary, somehow you cannot reduce your workers proportionately.

This is so because your automation has become one which sustains a fixed number of workers. During a time of stable growth, effort must be made to eliminate this "fixed number of workers" concept. Put your mind to creating a system which allows a line to produce flexibly a required number of cars with the number of workers according to need.

When this is accomplished, you can produce 70 percent of the production target with only 70 percent of people.

The first step toward "fewer people" requires our re-evaluating thoroughly the automation itself. In the process, we must free ourselves from the myth that automation equals a fixed number of workers.

"This operation can be automated, therefore we automate" is not an acceptable approach. First ask yourself this question: "Is there no other alternative but to automate?" For example, you have done all you can by way of work improvement, you have a fraction, e.g., 0.2 worker, left, and you want to reduce this to zero. When there is a need like this, perhaps automation is the solution.

The second step toward fewer people is to think about the placement of workers around a large automatic machine. Because of its size, workers assigned to that machine may be placed quite far apart from each other. Also, in most instances, the amount of work each must perform is only fragmentary, not requiring one full worker. Thus the thinking process may be trained in the following manner: "Can the work being done by worker A be transferred to worker B? And can the work of B also be transferred to C?"

In other words, attempt to make the work area of workers and the work area of a machine one and the same, if possible. This can then lead to "fewer people."

Figure 36. From People Saving to Fewer People

How to Attain Man-Hour Reduction

At Toyota, we promote man-hour reduction by first introducing the concept of time on hand, and then by changing work combinations.

Look now at Figure 37. There are six workers, A through F. The cycle time is set at one unit every three minutes. When work is performed on this basis, each worker acquires a little bit of time on hand. Transfer a little of B's work to A, and also transfer a little of

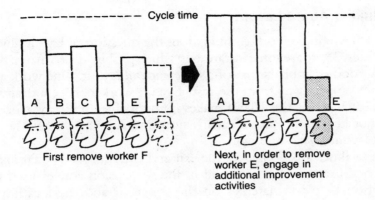

Cycle time

A B C D E F

First remove worker F

A B C D E

Next, in order to remove worker E, engage in additional improvement activities

Figure 37. Work Combination

C's work to A, until A's cycle time is completely filled. Repeat the same process with all the workers sequentially until each worker's cycle time is filled. If this is done, the work of F will disappear. So F is removed from this line.

In another illustration, a line consisting of A through E shows that for the one-minute cycle time, E has work for only 25 seconds. But the line remains a line for five workers. In this case, the task is to discover a way to remove 25 seconds from the entire line's work. Think hard, in order to remove this waste also.

For example, parts may be brought to the side of each worker, jigs may be improved to reduce the number of times they are passed from one hand to another or tools may be hung above each worker in the sequence they are used. With these changes it may be possible to shave off 25 seconds, and E can then be removed from the line.

However, if in spite of all the effort the 25 seconds cannot be removed, then extra tools may be brought in or some manual operations may be replaced by machine operations to come up with the 25 seconds. In this procedure, we follow our own maxim of progressing from work improvement to facility improvement.

If there are multiple machines, consider installing the buttons (switches) differently or in reverse order. The place where the button is located on each machine is important. For example, at the first process the worker places his work on the machine, and he walks over to the second process. A button must be placed in such a way that he can push it while walking. With this simple procedure, excess work may be eliminated.

Thinking About Layout

In arranging a work combination, the question of layout always surfaces. There are some instances in which problems also may arise.

A good example is that of a *remote island* layout. One work area stands alone, away from the rest. Even if workers have a desire to help each other, this layout makes it impossible. Similarly, workers cannot help each other if each is surrounded by machines. We call this layout *a bird in a cage*.

If facilities are elongated, the exit and entrance may be far apart. A good example can be found in the suspension conveyor. If the location to place materials on the conveyor and the location to remove them are separate, a minimum of two workers will be required. Our solution is to design the conveyor to make a turn, returning close to the place it started, and make its entrance and exit at the same location. There will be no stock on hand, and only one worker will be needed. In designing a layout, remember that the entrance and exit should be the same.

Problems may arise in the way we use the conveyor. Some conveyors are used exclusively to transport goods. For that reason, the entire layout may be extended: workers are placed one here, one there. It is again like the arrangement of remote islands, and the workers are not able to help each other. In cases like this, Toyota simply removes the conveyor.

Also, instead of having one conveyor line that can transport goods at a faster speed, Toyota prefers to have multiple lines that are shorter.

When a new layout is planned, it is important to observe whether it satisfies the three basic conditions of flow of goods, movement of people and flow of information. The most important consideration is that the layout support the system of orderly flow production. It is not desirable to have a layout that separates functional areas from one another, e.g., putting all the milling machines in one location.

Here are some pointers in planning a layout.

Put the entrance and exit together. First, when the entrance and exit to a process is made the same, one can engage in the practice of "when one part is withdrawn, another may enter." In this way, parts on hand in each process can remain constant. More importantly, this layout can inculcate in the minds of workers the importance of the just-in-time concept.

A second benefit is that the work area becomes clearly defined. Just as with automatic machines, if a worker is stationed at the entrance and exit of goods, it is possible that no one else is needed. By putting the entrace and exit together, a worker's work area becomes clearly defined. In this way, his work becomes more efficient.

Another benefit is that there is no wasted motion. Where processes are operated manually, the wasted motion of going to some other place and returning is eliminated. (Refer to Chapter 3 for an example of processing a certain gear.)

Finally, it becomes possible to have fewer people consistent with the amount of work. To put the entrace and exit together means, in practical terms, to create a layout with a square bracket shape, a U shape or a round shape. In this way, as discussed above, wasted motion is eliminated and, depending on the amount of work required within these differently shaped layouts, workers can be added or reduced.

A single long line process is what we might call a "go away" layout and is not desirable. The above explanation fully demonstrates why.

Concentrate work areas. No further explanation is necessary here, except to note that this issue is fully treated in the section on how to attain man-hour reduction. We know, of course, that "remote islands" and "a bird in a cage" are undesirable.

Use a people-centered layout. We often see in factories that power units and control units are placed right in the middle of the workplace. They hinder people's movement. Perhaps these were installed before questions of layout ever became an issue. But correct these mistakes if you can, and do so immediately.

Do not let one motor power multiple lines. Just because a motor has excess capacity does not mean that it should be allowed to power both line A and line B. Suppose A is burned; so will B. Or when you want to operate only A, and B is not needed, line B still has to be moved.

Use wide frontage and narrow depth for your store. Under the kanban system, the store may hold many kinds of items, but the quantity for each need not be large. In principle, the store's layout should be one of wider frontage and narrower depth.

Implementation of Improvement

When a plan for improvement is approved and is about to be implemented, one suddenly discovers that the result cannot be known unless one first acts on the plan. The plan calls for eliminating waste and reassigning to three workers the work done previously by four. But when the work is actually redistributed to the three workers, there still remains 0.1 person's work. Can the supervisor insist that the three take over this extra 0.1 person's responsibility? It may incite resentment from the workers and contribute to union militancy. But one cannot simply abandon the goal because of these apparent difficulties.

Results are important for an improvement activity. Once started, be patient and work until there is a reduction in the number of workers. If your assignment is to reduce the work load by 15 seconds, consider one or more of the following possibilities: move the parts storage area closer to the work area to reduce the time needed to walk, make the pallet smaller, transform a push-button switch into a one-touch switch, automate the process of pulling products from the shelf or hang tools above the worker. Focus your attention carefully on only a few items and continue to think about them. A certain little hint now and then can go a long way in germinating superb ideas.

Another important issue is that it is necessary for the improvement plan to be stable. Those situations that have been improved must be incorporated into your standards as new ways of doing things. If all the improvement measures are merely ad hoc in nature, they wil be of little use to the company. When improvements are made in facilities, jigs, tools and chutes, observe them constantly until the workers have mastered the new ways of handling them. In performing standard operations for exchanging cutting tools and dies, do them thoroughly until the workers can master the entire procedure. Do not be complacent because you have the standards; they may be imperfect at times, and part of your work is to improve them.

Without cooperation from the workers, no improvement plan can succeed. To obtain the cooperation and understanding from the workers, pay attention to the following two points:

1. *The worker must know when it is a spare moment.* Ask the workers who have time on hand not to do anything during that time. For example, the line turns with the one-minute cycle time. A worker completes his work in 40 seconds. For the next 20 seconds, he must stand there idly. Thus everyone, including the worker himself, will know that he has a spare moment. When the time comes to increase his work load, there will be no resistance.

2. *When you have to make a cut in your personnel, cut those who are superior.* Whom do you remove from your line? Do you remove those who are superior? Or do you remove those who are unskilled, difficult to work with or unaccustomed to the work? Many managers tend to do the latter, but when they do, the one who is removed will never grow in skills and may become resentful of others. It is not good for the morale of the workplace. Conversely, when those who are superior are the first ones to be removed, workers may show a positive, cooperative attitude.

8

Producing High Quality
with Safety

True Worth of Improvement Is in Quality

The major aim that transcends everything else for a manufacturing industry is to create products of high quality.

No matter how large a quantity a company may produce, if the product is of poor quality, no customer will buy it. No matter how cheaply a product is made, if that cannot be translated into money, in the end it means a loss. In the case of an automobile, safety is very strongly emphasized. We cannot let go of quality because we are too busy or because we want to make our cars cheaper. To send that kind of product to market is anti-social. It can also become the very factor that could destroy the company.

In other words, securing quality is the company's first concern. If there are other considerations, and quality is overlooked, that constitutes a classic case of putting the cart before the horse.

Among the many operations we engage in, which are the ones that can secure quality?

In prior times, the horse sense and skill of experienced workers were determinant factors in product quality. In today's diverse processes which rely on the division of labor, quality is secured through standard operations, which are generally determined under the prevailing working conditions. In short, standard operations are determined in such a way as to secure the requisite quality. However, even with this, quality can vary widely, and visual inspection and checking with a gauge must be built in as processes within the standard operations.

If defectives are produced under these conditions, it can mean one of the following: that the workers are not following standard operations, or that there is a breakdown in any one of the machines, dies, jigs or tools. We shall discuss the first problem first.

Once in a while we hear that after implementing man-hour reduction activities, defectives actually increased, or that so many people were removed from the lines that its effect was felt as a reduction in quality. Neither of these should ever happen. As discussed earlier, according to the Toyota system, these are instances of putting the cart before the horse.

When we look at actual occurrences of defects in the workplace, they can be divided into the following two categories:

1. Within a fixed time period, a worker may feel that his work load has been unjustly increased, and he either omits the work he is supposed to do or forgets about it. In other words, instead of eliminating waste, he has been engaged in an act of omission.

2. Heretofore, it has been possible to store products in the intermediate storage area or to rework defective pieces, due to the excess in manpower. Poor quality which has not surfaced before suddenly becomes an issue because of the effectiveness of the man-hour reduction activities.

The first type of situation is often found in assembly lines that use a conveyor. This problem is caused by not stopping the line when work is delayed or when problems arise.

Omitting work because one cannot catch up with the work he is supposed to do represents a thinking that the line should not be stopped. It is far better to stop the line and ship to the subsequent process a product without defect. This view must be inculcated in the minds of workers by their supervisors.

Do not worry about the line speed or the cycle time. Everyone must be familiar with the important concept that *the cycle time has nothing to do with the number of people at work*. Ask each of your workers to complete his work cycle. It means that he does his work at his own pace, but must do everything that is required. If the worker cannot finish his work within the cycle time, stop the line until his work is completed. A question may arise as to whether the time required must be added to the cycle time. The line need not be concerned with this issue, and that task is squarely on the shoulders of managers, supervisors and engineers.

For example, the cycle time is set at 60 seconds, but one worker goes through processes 1 to 5 in 70 seconds, or 10 seconds above normal. Do not discard the 10 seconds while the processes are progressing. The worker must work at his normal pace. The line is to stop for 10 seconds each time, in order to allow this worker to produce a quality product.

However, it is the responsibility of supervisors and engineers to find ways to ensure that the work can be completed within 60 seconds, and at a normal pace. They can remove waste from each process and shorten the distance walked. Only after appropriate actions such as these are taken can there be an assurance that the line stop will cease to exist.

It is easy to order that no line be stopped. But if this is done without improvement in the operational processes, it invites unevenness in quality. We do not allow that to happen under the Toyota system.

Defects that have not been apparent become visible in the second case when a reduction in manpower and inventory occurs. Previously, defective goods might have appeared with regularity, but were corrected internally without any steps being taken toward a fundamental solution. For example, the following can fall within this category: defective items produced by the preceding process are corrected by the present process, without providing any feedback to the preceding process; or tap holes do not fit because of design defects, but the present process simply taps them to correct them. The true cause is left undiscovered because of these makeshift corrections. Man-hours and storage needed to accommodate these makeshift corrections raise the cost.

The best opportunity for improvement is when defectives become clearly identifiable through man-hour reduction. Supervisors and engineers must return defective goods to each of the sections responsible for them and thoroughly investigate the cause or causes. If necessary, they must go to the preceding process to do the same. They must find fundamental solutions to the problems thus raised. A doctor cannot treat a chronic case of appendicitis by cooling off the affected area. Appendectomy is the only way to restore the person to complete health. In our factories, a similar approach is required.

Our thinking extends to the solution offered when defectives are caused by machines, equipment, dies, jigs and tools. For example, if defectives are caused by equipment, stop the line immediately and eliminate the cause.

Do not rework defective items within your own process simply because those processes responsible for them do not respond easily. Once you start reworking, it becomes part of your regular process without your becoming conscious of it. Instead, ask the preceding process to correct them. Your action does not end with a memorandum or a phone call. You must patiently seek corrective measures until a truly high-quality product is manufactured.

Inspection Does Not Produce Added Value

If the final assembly line allows defectives to go through, there is a greater possibility of defective goods reaching the hands of customers. Normally, defectives are discovered through the process of inspection and are reworked before reaching the hands of customers. The stronger the desire not to ship defectives, the stricter the inspection. Reworking may also become more frequent. In this way, however, the cost also rises.

Inspection done by inspectors outside the regular process does not produce added value. Therefore, those who are actually engaged in manufacturing must be the ones responsible for full quality assurance. They must not allow defectives to be attached to their jigs, and they must always apply gauges to inspect. Try to do everything right the first time. In principle, reworking must be considered something that is not allowed to occur. Both reworking and inspection by outside inspectors add man-hours to the same products. The ratio of the factory's added value goes down while the cost rises.

Will your customer accept your contention that "this product has been inspected ten times, and that's why it is expensive"? Work that does not produce added value is a mere waste. One can eliminate waste in the actual processing and reduce man-hours. But if the end result is to produce defectives, it may simply mean the man-hour requirement for inspection and reworking is substantially increased. From the point of view of cost reduction, the net result may be a zero or even a negative one. It may stray far from the original goals.

This being so, eliminate as much as possible both inspection outside the regular process and reworking. They are actually a waste. Produce good products so that inspection and reworking become unnecessary. Man-hour reduction will follow naturally.

Quality Is Built In by the Process

Previously the practice was to let inspectors inspect the parts produced and then let the subsequent process receive them. However, once these parts were manufactured, the act of passing judgment on them to say good or bad did not result in creating good-quality products.

An inspector may pronounce a product to be good after engaging in a sampling inspection. But if there is one defective product among many thousands, the customer who purchases it is not going to say, "This is one lemon among thousands of good cars. Too bad I'm the one stuck with it." For this reason, it is necessary to think in terms of inspecting everything in one form or another.

Hence, a view is born that full-time inspectors are to be eliminated, accompanied by the notion that quality must be built in by the process itself.

To build in quality means that each worker is responsible for each of the work processes he does, and must ascertain quality for each. Inspection must be brought within the process, and in order to send only good products to the subsequent process, defectives must be picked up on the spot. Our slogan is: "Catch the defective in its act." It is imperative that workers check their own work and subject every piece to their own full inspection. The next process is their customer. Defective items are never to reach the subsequent process. This is the key to our approach that quality is built in by the process.

Of course, the methods of inspection must be carefully investigated. Along with visual inspection and use of gauges, foolproofing or "avoidance of unintentional mistakes" must also be considered.*

If work is done by lot on a high-speed automatic stamping

---

* See *Zero Quality Control: Source Inspection and the Poka-yoke System,* by Shigeo Shingo (Productivity Press, 1986), in which a full discussion of Toyota's foolproofing techniques is given.

machine, retain 50 or 100 pieces over the chute, and inspect the first and the last piece. If both of them are good, then move the entire lot to the pallet. If the last piece is defective, find out where the defect began and remove all the defectives. At the same time, take steps to avoid recurrence. This is, in a sense, an inspection system which calls for 100 percent inspection. Never think in terms of sampling inspection just because the machine is a high-speed one.

Even after this, if the subsequent process discovers a defective part, it must immediately relay that message to the preceding process. The process that has received this information must stop its processing activities, probe into the cause and take action. Whenever a defective is discovered, communicate that fact immediately to those responsible for it. If this is not done, defectives will be produced continuously.

Reworking of the defective products must be done only by workers at the process responsible for them. Never say, "Oh, it's nothing," and quietly make the correction at the next process. That is one of the ways to perpetuate defectives. The division or process responsible for the defectives must be the one to correct them.

Don't Issue a Death Certificate

Now let us consider the issue of inspection done by inspectors.

A common idea is that inspectors differentiate between good and bad products, add up the results and send a report to the preceding process. This is usually considered to be the end of inspection — but this perception is insufficient. Inspectors must consider themselves staff members who are given the responsibility to analyze the reasons for the occurrence of defectives, probe into the causes and let the practice end. They are not examiners whose only function is to assign passing and failing grades. They must be able to explain to the workers why mistakes took place and teach them not to make the same mistakes again. Their function is analogous to that of a private tutor.

When wrong assembly of parts occurs, often the inattentiveness of a worker is cited as the cause. But the matter may not be that simple. It may be that the parts were not properly lined up in the order they were to be assembled, or that the line stop button or the call button was too far way, or that the work directions bulletin was not easy to read. There are likely to be a number of causes. Only by knowing what the causes are and taking appropriate actions can one take a step toward reducing the number of defectives.

The goal that an inspector must set for himself is not to throw away defectives, but to eliminate totally their production. This must be established as the criterion for judging his performance.

Foolproofing (*Poka-yoke*)

To have quality built in by the process, what must the workers do? What points do they check and which parts do they measure? When are the cutting tools to be exchanged?

Consider what roles the jigs, tools and mounting tools can perform in helping solve these problems. They can be made to inspect automatically the products received from the preceding process. You can foolproof your process and uncover defectives.

The process of foolproofing (*poka-yoke*) must be standardized to ensure that stable quality can be assured with a minimum number of man-hours, even when another shift comes in.

Without intending to, and no matter how careful he may be, a person can make mistakes when taking a measurement or checking products item by item. A way must be found to prevent the occurrence of disorders such as producing defectives, taking missteps in work processes and sustaining injuries. The foolproofing system we suggest is a means to create devices that can discover disorders without the workers having to be attentive to minute details. Among the devices to be considered are the following:

- If there is a misstep, the device does not allow goods to be mounted to jigs
- If a disorder is found in the goods, the device does not allow the machine to start processing
- If there is a misstep, the device does not allow the machine to start processing
- If there is a misstep in work process or in motion, it is automatically adjusted, and the device will allow the processing to proceed
- The disorder that has occurred in the preceding process is examined at the next process, and the device will stop defectives

- If a certain operation is forgotten or skipped, the device does not allow the next process to begin.

Methods to be considered in foolproofing include the following:

1. **Display method:** Light a lamp, making it easier to recognize by color and the like. This is a visual control method that makes it easier to discover disorder with the eyes.

2. **Jig method:** Foreign matter cannot be mounted, or when a mounting miss occurs, nothing can be moved. In this method, jigs are adapted to the task of discovering disorders.

3. **Automatic method:** The machine stops if a disorder occurs while processing. Some people do not include this method as part of our foolproofing system.

In establishing a foolproofing system for your company, select those areas which are easy to manage and in which the least amount of loss will occur.

Safety Transcends Everything

There is a saying that "spilled water never returns to the bowl." In our human existence, there are some things that can never return to their original state. Machines and other facilities may be restored after a breakdown if a sufficient amount of money is spent. But if injury occurs to people, their bodies may never be restored to their original vigor. There can also be deaths resulting from accidents, and no money can buy back a life. Safety is always our first and foremost concern, and there can be no man-hour reduction activity without consideration for safety.

Every method available for man-hour reduction to reduce cost must, of course, be pursued vigorously; but we must never forget that safety is the foundation of all of our activities. There are times when improvement activities do not proceed in the name of safety. In such instances, return to the starting point and take another look at the purpose of that operation. Never be satisfied with inaction. Question and redefine your purpose to attain progress.

Safety and man-hour reduction may at first glance appear to be contradictory, but the approaches of the two are actually identical. This is so because man-hour reduction also promotes elimination of waste (*muda*), unevenness (*mura*) and unreasonableness (*muri*).

In every manufacturing factory, most accidents occur due to movement caused by waste, unevenness and unreasonableness *(muda, mura* and *muri)*. In short, the factory may be forcing their workers to do what they need not do, or what is very difficult to do. Their movements thus become one of waste, unevenness and unreasonableness, resulting in injuries to the workers.

In Japanese, the word for injury is *kega*, which consists of two Chinese characters, *ke* and *ga*. If one reads this word *kega* closely, it means that "even for myself *(ga)*, it looks strange *(ke)*." Movements that are strange, e.g., with waste, unevenness and unreasonableness, are the causes for injury. Eliminating waste, unevenness and unreasonableness is directly connected with safety.

Generally, injuries occur in workplaces where there is inadequate daily maintenance. Putting away things that are not needed, putting things in readiness for use, maintaining cleanliness, engaging in cleaning and establishing work sequence and pointers are among the concerns that any self-respecting factory must have. But in workplaces where these concerns are not discernible, accidents occur frequently. Conversely, in workplaces where activities for man-hour reduction and improvement are carried out vigorously, accidents seldom occur.

Procedures in the workplace must be simplified. The greater the degree of simplification, the easier it is to manage and discover abnormalities.

As to the work motion, simplification is also needed.

Simplified work motions have hardly any waste, unevenness and unreasonableness. They are easier to handle and manage. Simplification reduces the chance of having unsafe activities. It is important to institute a system of visual control through which motions that are unsafe or unstable can be promptly discovered.

We have promoted simplification in our man-hour reduction activities. As can be seen from this discussion, simplification is also an important consideration in safety.

Putting away things that are not needed, putting things in readiness for use, maintaining cleanliness and engaging in cleaning are four activities that are conducted on the basis of the existing personnel and equipment. Therefore, if there are too many people or goods, or if the layout of machines is poor, the four activities will not work effectively. If the process is moving in the direction of complexity, there is a limit to the movement for putting away things. We must not overlook the fact that in the workplace the relationships between people, goods and facilities are intertwined, and no one is independent from another. For example, if there are too many people, too much can be produced (stockpiling). This, in turn, calls for people to put things away, pull things out and put things back, engage in maintenance and do reworking — which, in turn, calls for even more people. Thus a vicious cycle is created.

If a workplace increases something, other things will also increase. Complexity grows in the exact proportion to an increase in number. Man-hour reduction is an important step in preventing the growth of this trend toward complexity. It is also an important consideration in promoting safety.

SAYINGS OF OHNO

The Japanese word seiri *refers to putting away those things which one does not need. The word* seiton *means to put things in readiness for use at any time. Just to put things neatly in line is called* seiretsu. *Managing a workplace requires* seiri *and* seiton.

To create a safe working environment, a company-wide commitment is needed. The first step toward that is to create a workplace that has no waste, unevenness and unreasonableness. To do so, we must create an environment that makes it easier to discover waste, unevenness and unreasonableness.

The following are the questions that must be incorporated and categorized in the manual on work directions:

• Under what conditions are the operations conducted?
• In what order?
• And in what time frame?

The conclusion is simple: promoting man-hour reduction can lead to safety. Thus, to create a safe working environment, man-hour reduction activities must be actively pursued.

Easy Automation Leads to Injuries

When automation lacks a human touch — if it installs equipment just for the sake of labor saving and not for people saving — it generally requires a worker to serve as watchman. Without this watchman, the machine may not function properly. In Toyota's automation-with-a-human-touch system, the automatic stopping device comes into play in an emergency. From the point of view of safety and of man-hours, this is an important factor.

A problem fairly common in automated factories can be illustrated by the following case, which actually happened in a machine shop.

In a process represented by Figure 38, a worker had his finger cut off by a roller. Normally, the worker's function consisted of packing the pieces produced automatically by the machine. There are only ten lines similar to the one illustrated, and normally that is the work load for one person. However, in the case under discussion, there were actually three to four people constantly making rounds of the lines. They had to make these lines flow smoothly or they could not meet their quota. The chutes did not flow smoothly. Switches that were to signify "no work," "passage OK," and "full work" were not set correctly, and machines would not stop in an emergency. The accident occurred because the finger was caught in a machine operated automatically.

In this particular instance, questions should have been raised initially to see if the work could have been performed by one person. If a person who could bring about improvement in the workplace had been there, he could have improved the flow of the chutes. He could have effected a true man-hour reduction with safety, and the accident would have been prevented.

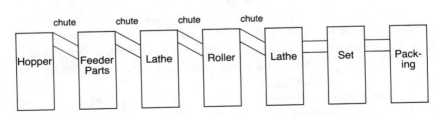

Figure 38. Easy Automation Leads to Injuries

Is It Dangerous to Have One-Touch Start-Up?

At Toyota, we are beginning to use the one-touch start-up method for metal-stamping and other machines. In most machine shops, this practice has been in existence for some time. They have always had the notion that one person can operate a number of machines. They have thought about this carefully, and they have been confident of safety in their operation.

The merit of having one person handle a number of machines is lost if the start-up of a metal-stamping machine calls for employing both hands to push the lower dead point. This is because the time spent is wasted. In addition, we must also add the time needed to walk from one machine to another. Why is it necessary to use both hands to push the lower dead point? It is merely to follow the provisions of Japan's Labor Health and Safety Law. The law requires that "machines such as metal-stamping machines must make provisions to ensure that no part of the body of a worker enters the danger area while slides or cutting tools are in motion. However, this provision shall not apply if the stamping machines have devices which stop the slides and cutting tools promptly when part of a body enters the danger area." The both-hand push-button device conforms to part of this provision, but in letter only. The device is not satisfactory. It is of no value to a third party, unless the operator of the machine is aware of his presence.

The one-touch start-up method is not wrong. But structurally the machine is still inadequate, because when part of a person's body enters the danger zone, there is no adequate device to stop the machine.

It is perfectly safe to have a one-touch start-up button if the machine contains a device which stops it immediately when a part of a worker's body enters the danger zone, or if there is another device which will stop the machine (which is started by the start button) if a person comes too near it. But these safety devices must not be operative when the machine itself is broken down. Also it will be better if a worker's body does not enter any part of the machine while working.

This approach has been gradually accepted in the metal-stamping and automatic welding lines. In one example, wire mats are spread on the ground under a 250-ton class metal-stamping machine; if anything touches the wires, a limit switch is turned on to stop the stamping machine. Or a stool is put by the side of an automatic welding machine; as long as a person stands on the stool, the machine does not operate. In both instances, the push buttons have been changed to the one-touch system.

These are still preliminary devices. If they can be improved and made applicable to others, all buttons can be changed into the one-touch system.

When we accept this line of thinking, we discover that there are so many safer and more rational ways of doing things when we return to the basics. How often do we forget this and feel that for the sake of safety we must foresake rationalization of work?

Figure 39 provides a good example of a positive approach.

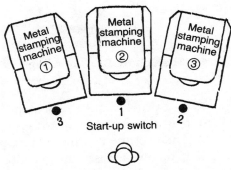

Figure 39. When One Worker Handles Three Metal-Stamping Machines

Roles of Supervisors at the Workplace

It is said: "A 10-percent cost cut is the equivalent of a 100% increase in sales." Cost reduction activities are extremely important for any industry. If these activities are neglected, the supporting pillars will be dislodged from their very foundation.

Supervisors who promote these activities make significant contributions to the company. Recognizing that, the Toyota system provides an analytical tool as shown in Figure 40.

In turning this control cycle, how can a supervisor implement the man-hour reduction activities which are consistent with the aims of the company? With what frame of mind and with what action?

There are two basic functions for a supervisor. One is to secure

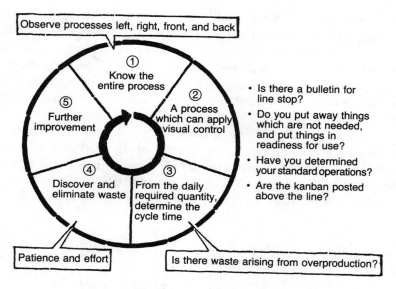

Figure 40. The Control Cycle of a Supervisor

the quantity required and provide quality assurance. The other is to engage in improvement activities for man-hour reduction.

The two basic roles contain elements which may appear contradictory, since on the one hand a supervisor must assure both quantity and quality, and on the other he must do so with the minimum of workers and equipment. The two tasks are not easy. If the supervisor places too much emphasis on securing quantity, he may not allow the line to stop. This may lead to an increase in the number of workers, in equipment and in storage. It means that costs will also rise, and his efforts end up in thwarting the company's aims. The supervisor must therefore learn to move his line forward in the direction that both requirements can be satisfied.

Abnormality Control

How can a supervisor manufacture products which can satisfy quality, quantity and cost needs? What are the concrete means available to him to improve his line?

There are many things that can become objects of control by the supervisor. To name just a few of them, they are: the workers, work assignments, teaching workers how to do things, achieving and changing quality and production plans, equipment, safety, storage, material preparation and setup. While a variety of control methods are available, if he tries to do everything, he needs more than a hundred hours in a day.

Under the Toyota system, everything is standardized, and the system emphasizes only those things that vary from the established standards. In other words, we teach supervisors how to engage in *abnormality control*.

At work, standardization means to determine standard operations and have everyone abide by them. In terms of materials and storage, standardization means providing clear-cut directions for the amount and place of storage. In terms of start-up instructions, it means the use of kanban; in terms of safety, it means establishing handling standards. Once these rules are established, workers must abide by them.

Once the workplace is organized in this manner, the supervisor can pay closer attention to those matters that do not fall within the rules. They can be taken up as problems to be solved. The more he follows this approach, the clearer he becomes in knowing what he must do. He reaches his control point, knowing clearly the duties of a supervisor.

What the supervisor must do first is organize his line neatly. In concrete terms, it means to establish categories of standards, assign places to store materials and parts, determine the quantity, establish kanban, and install call buttons, stop buttons and display lamps (*andon*).

After showing the workers the rules, which clearly indicate his intention, the supervisor must observe how these rules are to be carried out and what impact their operations may have on the workplace. If there are unexpected happenings which do not follow his professed intent, he may wish to revise his standards.

At this juncture, an important consideration is to be able to differentiate between those phenomena that are normal and those that are abnormal. In spite of the careful standardization, if abnormalities cannot be found or if after finding them he acts as if they do not exist, then the supervisor has failed in his responsibility.

Make sure that everyone can understand and think about what is abnormal. The eyes that have caught abnormalities have taken the first solid step toward improvement.

Here are some sample problems and sample solutions.

Worker A seems to have time on hand. This occurs every day. There may be a pile of parts which he has produced and stored behind him, or he may begin to do work not required under standard operations. All of these are abnormalities. Worker A does not have enough work to do.

Worker B, on the other hand, cannot finish his work on time. He stops the line. He skips some work and produces defectives. These are also abnormalities. B has too much to do. If A can do B's work within the time allotted, it means that the supervisor has not taught B well.

Worker C puts his hands on automated devices to assist in the process (which is, however, outside the standard operations). This is also abnormal. When the matter is investigated, it is found that jigs are unsteady and, if he keeps his hands off, defectives are produced. The supervisor must immediately call an engineer or maintenance to correct it.

Behind the line, there is a product without a kanban. This is also abnormal. One of the following reasons probably accounts for it: the cycle time is set wrong and the product comes off the line too fast, all workers have too much time on hand or the subsequent process has stopped withdrawing because something has happened there. In any event, it is a serious occurrence which must not be overlooked.

For the supervisor who establishes the standard operations, these are all abnormalities. The causes, however, are not difficult to find, and they can be one of the following: standard operations are unreasonably difficult, there are defective parts and materials or facilities are inadequate.

When the line is stopped, or when defectives are produced, it is easy to find that abnormalities have occurred, since they are all directly related to quality and quantity. But there are other abnormalities, which are caused by a little waste here and there, or by a tiny infraction of standard operations. They are easy to overlook, and may be considered of only secondary importance, but they are still culprits contributing to the company's higher cost. All of these abnormalities must be treated as important indicators of how the improvement and cost reduction activities must progress. No matter how small an abnormality may be, do not overlook it.

The role of a supervisor is to turn the cycle consisting of: standardization → discovery of abnormalities → probing for causes → improvement → standardization. By turning the cycle over and over again, he is able to fulfill the seemingly inconsistent functions of securing quantity and quality, on the one hand, and reducing cost, on the other.

When You Are a Supervisor

There are several necessary conditions for performing the task of supervisor adequately.

The first is always to *observe what is going on in the workplace*. If a supervisor does not visit the line and shows no interest in what is going on out there, that alone marks him as a failure; with this he is saying that he cannot check the standards which he himself has established. Of course, he will not be able to differentiate between what is normal and what is abnormal. How can he be expected to engage in improvement activities?

The second is to *control and guide his subordinates well*. This means to let them do what he wants them to do and train them toward that end. It is not to make the subordinates feel good, or even to defer to them. That is not a good way to establish a good human relationship. One of the subordinates will one day become his successor as supervisor. He must teach them, train them and in the end make them worthy of becoming supervisor. Only in this way can the present supervisor become the true "father" to all.

Under the Toyota way of thinking, we often ask this question: "Is this inventory ordered by someone, or is it created on its own?" We have materials, facilities and workers. But there is a time that nothing must be produced. The supervisor must be able to tell his subordinates when to stop and when to start. That ability to guide and control is an essential part of being a supervisor.

The third necessary condition is to *have a broad perspective and render judgment beneficial to the company as a whole*. No matter how effective a step may be in improving one's process, if that step is going to affect the preceding process or the subsequent process adversely, or to necessitate sending difficult work to outside contractors, then it cannot be considered a step toward improvement.

Every supervisor of a line must consider himself the manager of the line. He must have a broad perspective, ready to think in terms of the company's overall benefit.

After having done standardization and improvement, a supervisor who can say "please pull me out of this job" is by far the best supervisor. He has "trained" the line so it can move without him!

The Supervisor Is Almightly

Often we hear this question: "Is it good or bad for a supervisor to enter a line?" The answer given by the Toyota system is: "It is not good to be at the line at all times, but it is not right if he is not there at all."

If a supervisor is at the line at all times, he is no different from a regular worker. He will not be able to do the important tasks of managing, making improvement and providing education and training. Even so, he can never be totally free from the line, unless there is an overabundance of personnel. If he has been steadfastly promoting the waste elimination activity, it is natural that there are times he must enter the line.

He must never feel that he is forced to enter the line. The improvement activity for man-hour reduction is one of his most important

Figure 41. The Supervisor Is Almighty

tasks. The supervisor must leave line operations to become familiar with the overall picture. He does so because he has to engage in improvement activities. To promote improvement, he must know how easy or how difficult a given task is, and he must be thoroughly familiar with every procedure. There are always waste, unevenness and unreasonableness *(muda, mura* and *muri)* that cannot be known by just looking.

To teach subordinates, change the work sequence in the manner he sees fit and discover even the most minute waste, unevenness and unreasonableness, he must at times enter the line. When a worker is absent, it can provide the supervisor with an opportunity to enter the line to assist and to observe matters discretely.

The attitude that the supervisor assumes has an important bearing when he enters the line. Does he feel that he is forced to do so? Does he want to discover matters that must be improved? This difference in attitude becomes a crucial difference in the end. A supervisor who looks at his own action in the line positively can effect an improvement. He can also prevent work from becoming disjointed when a veteran worker is absent.

It is therefore not out of bounds to suggest to the supervisors that when an opportunity presents itself, they must enter the line. They must participate in its work with a forward-looking attitude of always wanting to learn and to improve. After all, management begins at the workplace.

Index

Abnormality control
  cowboys, by, 77-78
  sample problems and solutions for,
    156-157
  Toyota Motors, at, 78
Accidents
  causes of, 149
  kega, Japanese term for, 149
Added value
  inspection does not produce it, 144
  operations enhancing it, 126
  operations without it, 126
  source of profit in manufacturing
    process, 4
Andon, 155
  description of, 27, 74-75
  see also display lamp
Assembly line
  just-in-time applied to, 69
  load-smoothing production, 54
  production information by the
    minute, 81-82
Automation
  automatic machine, 71
  ji-do-ka, Japanese term for, 72
  just-in-time, and, 65-78
  labor saving, and injuries, 151-152
  not equal to a fixed number of
    workers, 133
  worker reduction, through, 132

Automation with a human touch
  ninben no tsuita jidoka, Japanese
    term for, x, 70
  outline of, 25
  Toyota system, in, 11, 24, 69-70, 151
Automobile industry
  manufacturing processes are linked
    in, 48-49
  production equalizing system in, 49
  production plan, changes in, 67-68
  quantities and types equalized, 49-50

Communication
  assembly line, 82-83
Control
  abnormality control, 77-78
  supervisor role in, 154
Conveyor, 11
Corolla, 8
Corona, 8, 47, 49-50, 51, 52
Cost
  elements of, 5-7
  manufacturing method determines it,
    7
  materials cost, 6
  personnel cost, 5-6
  "true cost," 5, 7
  waste as element in, 6
Cost principle, 7

161

Shingo, Shigeo (cont.)
A Revolution in Manufacturing: The SMED System, 58
Study of Toyota Production System, vi
Zero Quality Control: Source Inspection and the Poka-yoke System, 145
Shojinka
people saving system at Toyota, 132
Staff reduction
superior personnel should be removed from the line, 139
Standard operations
bulletin, 105
components of, 101-105
cycle time as component of, 102-103
implementation of, 117-118
improvement requires change of, 118
load smoothing as aid in, 56
methods of determining, 105-106
standard stock on hand, as component of, 104-105
work procedure, as component of, 103-104
workplace, in, 155
Standard Operations Bulletin
tells workers what supervisor expects of them, 111
Standard Operations Routine Sheet
form used at Toyota Motors, 107-109
Standard stock on hand
changes in, 104
definition of, 104
standard operations component, as, 104-105
Stock on hand
waste arising from, 17
Store
storage area at Toyota Motors, 95
Strokes per hour (SPH)
productivity measurement yardstick, 37
Supermarkets, xii
Structure applied to manufacturing plant, 65-66
Supervisors
abnormality control, 154

Supervisors (cont.)
control cycle of, 154
control responsibility of, 154-155
defective goods, returning, 143
line responsibilites of, 158-159
man-hour reduction, responsibility for, 154
performance measurement, 157-158
standard operations, role in, 117-118
workplace, role in, 159

Table of Part-Production Capacity
form used at Toyota Motors, 105-106
Tact *see* cycle time
Teamwork
overlap in workspace needed for, 115
Tenuki
omission, Japanese term for, 13
Time on hand, 16
waste arising from, 16-17
Tourist spots
peaks and valleys of work, 46
Toyoda, Kiichiro
term *just-in-time* coined by, ix, 66
Toyoda, Sakichi
founder of Toyota Motors, 70
Toyota Motors, v, vi
abnormality control at, 77-78
automation with a human touch system, 151
cost determined by method, 7
early days of the company, 119
foolproofing technique, at, 145-146
full-work system at, 96
have time on hand, 76
kanban system at, 56-57
kanban types used at, 84-85
lead time at, 41
man-hour reduction movement, 13, 134
one-shot exchange of die at, 63
one-touch start-up method at, 152-153
problem solving technique at, 27
production at, 8
production plan at, 79-80
ratio of operation at, 39
respect for humanity at, 13, 34, 131

Also from Productivity

Productivity Press publishes and distributes materials in the field of productivity and quality improvement for business and industry, academia, and the general market. Products include books, newsletters, audio cassette tapes, and audio-visual training programs. Productivity also sponsors training programs, conferences, seminars, and industrial study missions to Japan. Call for details, or send for our free brochure and book catalog.

Managerial Engineering: Techniques for Improving Quality and Productivity in the Workplace, by Ryuji Fukuda

A proven path to managerial success, based on reliable methods developed by one of Japan's leading productivity experts and winner of the prestigious Deming prize.

Dr. W. Edwards Deming, world-famous consultant in statistical studies, says of the book, "Provides an excellent and clear description of the devotion and methods of Japanese management to continual improvement of quality, knowing well that as efforts succeed in improvement of quality, productivity improves, costs go down."

ISBN 0-915299-09-7   **$34.50***

Manager Revolution! A Guide to Survival in Today's Changing Workplace, by Yoshio Hatakeyama

A landmark bestseller in Japan that explains clearly, concisely, and in depth the fundamentals of a new approach to management. Both the occupational and human aspects of the manager's daily job are closely examined. The author successfully blends theory with practice by including case histories and checklists with each chapter.

Says Philip Crosby, renowned management expert, "I wish I had had *Manager Revolution!* when I was beginning my management career. It is not so much about the techniques of running an operation as it is about leaving a lasting result from one's efforts."

ISBN 0-915299-10-0   **$24.95***

* Please add $2 for shipping and handling.

A Revolution in Manufacturing: The SMED System
by Shigeo Shingo

Shingo, a creator of the Toyota production system, presents an introduction to SMED (Single-Minute Exchange of Die), an essential component of the famous Just-In-Time production system used first at Toyota and increasingly in the United States.

Containing hundreds of illustrations and photographs, the book provides the most complete and detailed instructions available for transforming a manufacturing environment to speed up the production flow and make small-lot inventories feasible.

ISBN 0-915299-03-8   **$65.00***

Zero Quality Control: Source Inspection and the Poka-yoke System, by Shigeo Shingo

A companion volume to *A Revolution in Manufacturing: The SMED System*, this book proves that "defects = 0" is absolutely possible. This is truly a revolution in quality control.

Shingo's premise is that inspections and SQC do not reduce defects. It is really source inspections and *poka-yoke* (mistake-proofing) that leads to the complete elimination of defects. These are explained in detail, and over 100 specific examples demonstrate the wide range of applications for these systems. Essential reading for your quality improvement efforts.

ISBN 0-915299-07-0   **$65.00***

* Please add $2 for shipping and handling.

Productivity Press
P.O. Box 814
Cambridge, MA 02238
(617) 497-5146

Productivity, Inc.
P.O. Box 16722
Stamford, CT 06905
(203) 967-3500

PRODUCTIVITY

I would like to know more about the products and services of Productivity, Inc. Please send me information about the following:

☐ **The PRODUCTIVITY newsletter.** The # 1 monthly newsletter devoted exclusively to issues of productivity and quality improvement.

☐ **"Productivity the American Way" conferences.** Offered at key locations around the country, bringing together leaders in American productivity to share their ideas and experiences.

☐ **Regional and in-house seminars.** Our training materials, recently translated into English, bring the best Japan has to offer.

☐ **Industrial study missions.** Productivity in its most effective settings, here and abroad—especially Japan.

☐ **Audio-visual training programs.** Shigeo Shingo's techniques for Just-In-Time production, and the famous General Electric program on JIT.

☐ **The PRODUCTIVITY library.** Authoritative books and cassette tapes by the Productivity Press and selected books from other publishers.

Name _____

Title _____

Company _____

Address _____

City, State, Zip _____

Telephone _____

Productivity, Inc., P.O. Box 16722, Stamford, CT 06905   (203) 967-3500

DATE DUE

MAR 1 7 1995		5/28/08		
APR 1 4 1995 MAY 1 2 1995				
AUG 3 1 1995				
MAY 1 1 1996				
MAY 2 2 1998				
3/5/08				